ULTIMATE EXPLANATIONS OF THE UNIVERSE

Michael Heller

ULTIMATE EXPLANATIONS
OF THE UNIVERSE

Translated from the Polish by Teresa Bałuk-Ulewiczowa

 Springer

Michael Heller
ul. Powstańców Warszawy
13/94
33-110 Tarnów
Poland
mheller@wsd.tarnow.pl

Original Title: Ostateczne Wyjaśnienia Wszechświata
© TAIWPN UNIVERSITAS

ISBN 978-3-662-50207-5   ISBN 978-3-642-02103-9 (eBook)
DOI 10.1007/978-3-642-02103-9
Springer Heidelberg Dordrecht London New York

Cover design: deblik, Berlin

Printed on acid-free paper

Springer is part of Springer Science+Business Media (www.springer.com)

PREFACE

The longing to attain to the ultimate explanation lingers in the implications of every scientific theory, even in a fragmentary theory of one part or aspect of the world. For why should only that part, that aspect of the world be comprehensible? It is only a part or an aspect of an entirety, after all, and if that entirety should be unexplainable, then why should only a tiny fragment thereof lend itself to explanation? But consider the reverse: if a tiny part were to elude explanation, it would leave a gap, rip a chasm, in the understanding of the entirety. Every, even the smallest, success scored by science is a step in the right direction, a sort of promise that somewhere along that direction, maybe still a very far way off beyond a runaway horizon, lies the ultimate explanation.

Only rarely are such thoughts, or rather such moods allowed to come to light in the enunciations scientists make. But they linger in their sub-conscious, suppressed by declarations that all that scientists are interested in are the results of research; empty speculation they leave to philosophically-minded dreamers. However, as we know, a repressed sub-conscious gives rise to a variety of pathological conditions, and in the sphere of ideas pathologies are particularly dangerous. It could be said that by addressing the issue of ultimate explanations in science I have decided on a course of psychotherapy, above all for myself. The ideas allowed to stray into the margins of my scientific papers have finally to be brought to order, put down on paper and submitted to public discussion and assessment. In science there are no psychotherapies but only such that are collective in kind, and hence the presentation of my ideas in book form seems the best choice of a therapy.

In the first chapter I explain and discuss the schema of this book at length, thus I feel absolved from this duty in the Preface. I would only like to draw attention to the plural in the book's title: *Ultimate Explanations of the Universe*. If there are many of them, then the problem is still open.

I would like to express my deep gratitude to Teresa Bałuk-Ulewiczowa for translating my book into English, not only for maintaining a scrupulous fidelity in rendering my ideas but also proficiently reproducing the mood that attended them. I am likewise deeply grateful to Angela Lahee of Springer Verlag, thanks to whose professionalism and personal intuition throughout the entire process of this book's creation the work on it was more like a continuation of writing, rather than the struggle to smooth out style and sense usual in such situations.

April 2009

Michael Heller
Kraków, Poland

CONTENTS

CONTENTS

Chapter 12

TEGMARK'S EMBARRASSMENT 107

PART III

CREATION OF THE UNIVERSE 115

Chapter 13

THE DRIVE TO UNDERSTAND 117

Chapter 14

THE METAPHYSICS AND THEOLOGY
OF CREATION 123

Chapter 15

CREATION AND THE PERPETUITY OF THE UNIVERSE 133

Chapter 16

CONTROVERSIES OVER THE OMNIPOTENCE
OF GOD 139

CONTENTS

XI

CONTENTS

Chapter 1

~

ULTIMATE EXPLANATIONS

1. TO UNDERSTAND UNDERSTANDING

A powerful, not fully comprehensible instinct to understand lies dreaming in us. We would like to have everything fully comprehended, explained, and proved. So that there should be nothing left without its cause, moreover a cause that would remove all the anxiety of doubt and all the question marks. The grander the matter, the more we want to have it explained, to eradicate even the slightest suspicion that things might be otherwise. This longing for "ultimate explanations" is not fully comprehensible in itself, and when we want to understand it a nagging question inevitably arises: what does "to understand" mean?

In trying to answer that question philosophers of science have used up forests of paper and a sea of ink. In the age when Positivism was the prevalent trend attempts were made to dismiss all questions for which there were no answers in sight within the grasp of direct experimental methods, and it was postulated that the task of science was not to explain but to describe. But already by the classical period in the development of the philosophy of science on the basis of the methodology proposed by the Vienna Circle (Wiener Kreis) the question was not whether science explained anything, but what scientific explanation *meant*. If we agree that a description is a set of statements giving information about something, while an explanation is, in the most general sense, also a set of statements, but one that shows a logical connection between those statements, it is evident that, on the one hand, there is no clear line of demarcation between description and explanation, but – on the other hand – that explanation is something more than description.

1

M. Heller, *Ultimate Explanations of the Universe*, DOI 10.1007/978-3-642-02103-9_1,
© Springer-Verlag Berlin Heidelberg 2009

Even a cursory knowledge of any theory in physics is enough to realise that not only does it describe, but also that it explains. Or rather that it only describes once it has explained. The framework for every theory in physics is always a mathematical structure, usually an equation or set of equations, appropriately interpreted, that is with reference to the world. A mathematical structure is essentially a network of logical inferences appropriately encoded in symbols. To unlock and decode those inferences we have to resolve the mathematical structure computationally, usually by solving the equation or set of equations. The interpretation, in other words reference to the world, is not carried out directly, but by calculating the empirical predictions of the theory and comparing them with the results of what is actually observed. This procedure is in fact tantamount to inserting the results of experimental observation within the grid of inferences making up the theory's mathematical structure. Moreover, the world speaks only through the medium of mathematical structures. Experimental results are always expressed in terms of numbers, but these numbers are meaningless outside the structure containing them. The logic of the way the measuring apparatus works is essentially part of the logic of the mathematical structure serving as a framework for the particular physical theory. The structure of the given physical theory is as it were embodied in the construction of the apparatus.

A standard textbook of methodology envisages the distinction between description and explanation in the fact that a description is "a set of statements providing information on a particular aspect of reality without a clear reference of these statements to other statements," whereas an explanation is "a series of statements connected with each other by means of systematic proof."[1] If we accept these rather general expressions, then there is no such thing as a pure description in the theories of physics, a description is always an explanation. The same author writes that the distinction between description and explanation is of the same kind as between statements and proofs.[2] This formulation leads to the problem of ultimate explanations. In the process of proving something we cannot "go back to infinity," ultimately we have to adopt some axioms as the basis of our proof. Similarly in a description we have to start from a point of reference. Otherwise we risk moving back to infinity. So what does "the ultimate explanation" mean? We should find something that would be "an explanation for itself". Like God in Christian theology, who "Is because He Is" – the self-explanatory Absolute; His non-existence would mean a fundamental incongruity. But all the indications are that this Logic is not accessible to our reason, and if we want to rely on reason alone we must maintain a far-reaching respect with regard to theological rationale.

Obviousness, in the sense of something that is self-explanatory, turned out to be an embarrassment to the history of science long ago – from the time of the

Ptolemy–Copernicus controversy to the achievements of quantum physics. Our sense of the self-evident has grown up in the course of our encounter with the macroscopic environment and invariably fails us whenever we have to go beyond the borders of that environment. "Infinitely small" and "infinitely large" worlds are completely different from the one to which our eyes have become accustomed.

There is one more possibility: something in the nature of "circular explanations" – a closed chain of inferences: the current conclusion becomes the reason for the statements of which it is an inference. There are many ideologies which employ this sort of intuition to build up philosophical visions, but until a logical model is created showing that such an approach is not self-contradictory, they will only be visions and ideologies. Self-referential methods are applied fairly often in logic and mathematics, e.g. in the proof of Gödel's famous theorem, but such methods are still a long way off from what we would be inclined to call the ultimate explanation.

All these misgivings are not powerful enough to fetter the belief which is not only rooted deep inside our personal expectations but also resolutely set on the horizon of the ambitious human enterprise called science – the belief that everything has its reason. The endeavour to reach that horizon is the driving force behind science.

2. THE TOTALITARIANISM OF THE METHOD

At first sight the mathematical and experimental method of the contemporary sciences looks highly ascetic. Its very origins involved the withdrawal from overly complex metaphysical issues and a limitation to the analysis of the simple facts of experimental data. This constraint of the field of interest immediately brought a remarkable effectiveness. More and more phenomena, further and further away from ordinary experience, were subjected to the mathematical and experimental method, but the method itself continued to be interpreted very ascetically, within the bounds of the measurable. This approach gave rise to Positivism: whatever was beyond experiment was not worthy of scientific attention at all. With time experimental accessibility turned into a criterion determining existence. In its most radical phase Positivism tended to adopt a view that all that could not be verified experimentally simply did not exist. It is not difficult to spot an idiosyncratic kind of methodological totalitarianism lurking in this attitude. The mathematical and experimental method simply does not tolerate competition: whatever resists its application is annihilated. In its Neo-Positivist version this totalitarianism was reduced to a claim that the bounds of rationality coincided with the boundary of the mathematical and experimental method. Whatever lay beyond the confines of this method was beyond the reach of rationality and was therefore irrational, in other words bereft of a sense.

The consequence of such an attitude should be the conviction that the ultimate explanation of the universe lies within the grasp of the mathematical and experimental method. For if there are no explanations other than those obtained on the grounds of this method, then the ultimate explanation – the furthest-reaching explanation – must lie within its bounds. However, such notions were not voiced openly in the Positivist and Neo-Positivist era, since they did not concur with the Positivist principle of economy. The postulate of ultimate explanations smacked of metaphysics, which had been relegated, not only from the realm of science, but also from areas connected with science. Today, after the demise of classical Positivism, this attitude survives only in a few of the more radical groups of analytical philosophers. Many scientists, liberated from the Positivist scientistic straitjacket, are succumbing to the natural instinct to search for ultimate explanations, but are setting about it as it were "on an extension" of the mathematical and experimental method, not really worried by the fact that at a certain point on this quest the boundary between physics and metaphysics must inevitably be crossed. In general it is claimed that the right place for such endeavours is the popular scientific literature meant for the general public, while in the professional publications on their research scientists tend to refrain from embarking on philosophical deliberation. This is only part of the truth, since apart from overt forays into philosophy there are also a variety of highways and byways on which philosophy may creep into scientific research. One of them is the pursuit of theories and models which offer a chance for an "ultimate explanation" in a version sensed, or even concocted, by the scientist in question. What is more, on closer scrutiny of the history of science it turns out that this is a strategy that worked well even in the heyday of Positivism.

The tendency to pursue "ultimate explanations" is inherent in the mathematical and experimental method in yet another way (and another sense). Whenever the scientist faces a challenging problem, the scientific method requires him never to give up, never seek an explanation outside the method. If we agree – at least on a working basis – to designate as the universe everything that is accessible to the mathematical and experimental method, then this methodological principle assumes the form of a postulate which in fact requires that *the universe be explained by the universe itself*. In this sense scientific explanations are "ultimate," since they do not admit of any other explanations except ones which are within the confines of the method.

However, we must emphasise that this postulate and the sense of "ultimacy" it implies have a purely methodological meaning, in other words they oblige the scientist to adopt an approach in his research *as if* other explanations were neither existent nor needed. It is a psychological fact that people who practise the scientific method for a long time develop a compulsive habit of endowing methodological rules with an ontological sense, in other words they become convinced that explanations which transcend the mathematical and experimental method are pseudo-explanations, since nothing exists beyond the reach of this method. This is evidently a path straight into the Positivist ideology, and if for any reason the scientist does not want to succumb to it, he has to "extend" the scientific method to cover all that he wishes to study. Today the latter tendency seems to have a goodly following of adherents. Many scientists are probably not at all aware of the need to distinguish between the "methodological order" and the "ontological order," and treat methodological rules as if they were ontological principles.

Let us return to the postulate to "explain the universe by means of the universe itself." The word "universe" in this expression, which sounds rather like a slogan, clearly indicates that the science in which the drive to seek "ultimate explanations" is the most manifest is cosmology – the science of the universe. On the one hand, cosmology, envisaged in a certain sense as dealing with a complete totality, has no chance at all of seeking any explanations beyond its own area of research; but on the other hand, for instance in its formulation of the question of the origins of cosmic evolution, it as it were imposes an extraneous perspective. Here the distinction between the "methodological order" and the "ontological order" turns out to be very useful. But no distinctions can remove the tension between the tendency to be rigorously economical in research methods, and the longing for full understanding. It is in the field of cosmology that the controversy over "ultimate explanations" is raging at its most excited and impassioned state.

3. MODELS

In 1983 Jim Hartle and Steven Hawking published a paper in which they proposed the later renowned model for "the quantum creation of the universe out of nothing."[3] Their principal aim was to unify the general theory of relativity, in other words Einstein's theory of gravitation, with quantum physics, making up a single, integrated theory of physics. In this paper they put forward an approximate scheme for the quantisation of gravitation and tried to show that within the framework of this scheme there existed a finite probability of the universe emerging in a certain state out of an "empty" state. They called this mechanism "the quantum creation of the universe out of nothing," which became the point of departure for many other papers and a sort of paradigmatic example of "the ultimate explanation in cosmology." We shall devote one of the following chapters to the Hartle-Hawking model.

One of Hawking's students, Wu Zhong Chao from China, was so fascinated by the Hartle-Hawking model, which Hawking and his students later continued to develop, that he dedicated a separate monograph to the subject, published in English in China.[4] In this book, particularly at the beginning of Chap. 3, Wu makes a number of comments of a methodological nature on the subject of "the ultimate explanation" in cosmology. They apply directly to Hawking's model, but in fact are more general in character and therefore deserve closer attention here.

Like all other theories in physics, cosmological theories must obey the same principles regarding proper method: above all they must be self-consistent, viz. not mutually contradictory from the logical point of view, and at least not contradictory with respect to the observed empirical facts. The former requirement must be kept rigorously, while adherence to the latter requirement allows of a certain degree of tolerance described at length in contemporary textbooks of the philosophy of science. The point is that sometimes it is better to have a theory which has problems with explaining certain "slight experimental discrepancies" than to have no theory at all. This was the situation in the late nineteenth century, for example, when it was already known that Newton's theory of gravitation failed to explain certain "small aberrations" in the orbit of Mercury (its perihelion motion), but the theory still continued to be used very successfully. As we know, it is not easy to codify all the rules of the methodology of science, but in the everyday situations of research common sense supported by tradition and experience tells the scientist the right way to proceed. Research in cosmology must obviously comply with these procedures.

But cosmology has its own specifics. Wu also requires cosmological theories to be self-contained. Let's explain what he means by this. Usually the mathematical backbone of any physical theory entails a differential equation or a set of differential equations. Not only is there a need for such an equation (or set of equations) to be solved, but also the initial or boundary conditions have to be determined (or "imposed by hand," as physicists say) for the selection of the right solution for the particular physical problem. For instance, if the equation is to describe the motion of a given body, then the initial conditions may be its position and velocity at the start of the motion. If the equation is to describe the gravitational field of a given star, then we may select the behaviour of that field at an appropriate distance away from the star as the boundary conditions for the right solution, e.g. we may assume that at infinity (viz. an appropriately long distance away) the field strength is negligibly small. In other words it is the researcher who selects the initial or boundary conditions on the basis of his understanding of the physical situation he intends to model.

We could, of course, do likewise in cosmology. A cosmological model is a set of differential equations, too, and when we select a particular solution to it we also have to decide on particular initial or boundary conditions. Usually we will be guided in our choice by the principle of simplicity, or we will try all the possibilities and adapt the equation to the experimental data ex post, or else (as sometimes happens) we build up a "philosophy" to the solution we have found. The point is that none of these options is applicable to the physical situation in cosmology. Initial or boundary conditions are extraneous to the model; they are something the physicist has to "insert manually" into the model. The universe is a physical system still undergoing a process of evolution which cosmology is successfully reconstructing, thus its initial or boundary conditions must have been fixed in one way or another. Only there was no hand to manually insert them into the world. Or – to put it more precisely from the methodological point of view – within the framework of the mathematical and experimental method we are not allowed to assume that such a hand existed. We have to do without its assistance.

In a nutshell, we should look for initial or boundary conditions outside the universe. But we may not speak of the universe not having an exterior; we should rather say that the concept of an exterior is meaningless with respect to the universe. Wu calls this logical loop the problem of the First Cause,[5] which has been the bane of cosmology ever since Newton's time. The theory of cosmology would be self-contained if it managed to disentangle itself from this problem. As we shall see in subsequent chapters, cosmologists have been searching for such a theory (or model) along various paths. For instance we may imagine a theory which would not require any initial or boundary conditions, or a model which automatically determined its own conditions. We shall also see that many authors have resorted to highly exotic ideas to make the world self-contained. This is undoubtedly a philosophical motive, but one that derives from the right methodological perception, and is today a very powerful trend in the thought on the universe and cosmology.

4. ANTHROPIC PRINCIPLES AND OTHER UNIVERSES

The goal of unity is firmly encoded in scientific method. Modern science started when giants like Copernicus, Galileo, Kepler and Newton managed to unite "earthly physics" and "celestial physics," in other words to show that the same laws of physics hold on Earth and in astronomy. For some time thereafter it seemed that the laws of mechanics discovered by Newton were the ultimate, "unified" theory governing everything. The discovery of electricity and magnetism finally swept this illusion away, but soon Maxwell showed that these two classes of phenomena could be put together in a single, mathematically very elegant theory of electromagnetism. Einstein was the first to come up with the idea that Maxwell's theory should be combined with the theory of gravitation, and devoted the rest of his life to the implementation of this idea. Today we know that Einstein's concept had no chance of success, since apart from electromagnetism and gravitation there are two other fundamental physical forces: the weak nuclear force (the lepton interaction), and the strong nuclear force (the hadron interaction). We now have an experimentally confirmed theory (the Weinberg-Salam theory) uniting the electromagnetic force with the weak nuclear force in a single interaction, known as the electroweak interaction. In principle we also know how to combine the strong nuclear force with this interaction. We have several scenarios for the unification and are only waiting for the experimental data which will select the right scenario. Only gravitation, by virtue of its different character, is defying the successful application of the unifying schema with respect to itself. No wonder that more and more tension is building up in the search for a quantum theory of gravitation – for it is almost certain that gravitation will have to be quantised before it can be unified with the other interactions. Virtually every new mathematical model, as a rule more sophisticated than its predecessors, at first generates enthusiasm and new hopes, but is soon relegated to the gallery of interesting but abortive constructions.

The multiplicity of unifying models put forward hitherto is splitting up physics into separate schools and trends rather than uniting it. This process of fragmentation has reached an apogee in the theory which probably the largest group of physicists regard as the most promising today – the M-theory, a development and generalisation of the superstring theory, which has become well-known outside physics. According to the good old quantum theory there should be one minimum energy state, that is the fundamental state. In M-theory there is "practically an infinite number" of fundamental states (estimated at even as many as 10^{500}). The problem is that it is the fundamental state that to a large extent determines the physics of the universe. What are we to do with such a vast number of fundamental states? Unless we reject all of the theory leading up to them, the only solution is to accept that there is a very large number of different universes – as many as there are possible fundamental states – each with its own, different physics. M-theory people speak of the "string landscape" of the various universes, and conduct research on it.

It was easier to accept such a situation psychologically, as the idea of many universes had been circulating for some time in discussions on certain issues in cosmology. It first appeared in connection with the anthropic principles, which in various ways formulated the observation that the existence of living organisms on at least one planet in the universe depended in a very sensitive manner on the universe's initial conditions and its other characteristic parameters. A slight change in any of these conditions or parameters in general gives rise to drastic changes in the universe's evolution, rendering the emergence of biological evolution impossible. For instance, a very small change in the universe's initial rate of expansion (either its slight acceleration or retardation) would preclude the emergence of carbon, the very cornerstone of organic chemistry. There are many such "coincidences."

What has made the universe "friendly to life"? This brings to mind the notion of a purposeful design of the universe. But such an idea is alien to the rule of "explaining the universe by means of the universe itself." To neutralise it, the following argument was employed: suppose there exists an infinite number of universes representing all the possible combinations of initial conditions and other parameters characteristic of the given universe. Only a very few of those universes are life-friendly, and we live in one of them, since we could not have come into existence in any other. Some of those taking part in the discussion immediately acknowledged the hypothesis of many universes as more rational than the hypothesis of the existence of God, while others said that there was no need to "proliferate existences" if a single God was enough.

Regardless of these theological disputes, the idea of a multiverse (as it soon came to be called) launched into a life of its own. Soon fairly concrete cosmological models, e.g. inflationary models or certain unifying scenarios, started to identify mechanisms which could have produced either different universes, completely separated from ours, or regions in our own universe to which we shall never have perceptive access. At any rate, the multiverse trend became a reality. But was it still science? Can something which we will never be able to access by observation, even in principle, still be a subject for scientific study? Or maybe the scientific method was being transformed before our very eyes, and something that was not science before was now turning into science? Nonetheless, I think we should not be too hasty in undermining scientific method – that's right: in rashly undermining the scientific method, which is rightly regarded as the greatest achievement of science, and on which all the other achievements of science depend. Instead we should once more recall that the boundaries of rationality do not coincide with the bounds of scientific method, and therefore it is sometimes worthwhile to transcend those boundaries in order to be able to carry on a rational discourse "once on the other side." Although we should not expect any empirical solutions in that area, critical argumentation and rational appraisal will still be relevant there.

5. CREATION OF THE UNIVERSE

The discussions on the anthropic principles and the multiverse are as it were on an extension of scientific research. On the whole it is quite difficult to identify the point at which we cross the boundary between what may still be called a cosmological model, and what definitely belongs to speculation beyond that boundary. But we could be even more daring and locate our observation point well beyond the boundary of scientific method (yet still within the area of rationality), and once we are on the other side take a look how the mathematical and experimental method works within the area proper to it, and what happens to its explanations as it approaches the limits of its possibilities. The area "beyond" is very well-known in the history of human thought: it is the region inhabited by philosophical and theological concepts. It is vast and highly "speculative." To prevent us from losing our way on its tortuous paths, I shall limit the area by applying two restrictions: First, in practice I shall not go beyond the concept of creation as rooted in Judaeo-Christian thought. The concept of creation undoubtedly entails the ambition to explain, though in a theological sense. It is a theological concept, but has acquired numerous philosophical accounts (in the light of diverse philosophical systems), and it is chiefly the philosophical aspect of the idea which will be our subject of study. I shall touch on other philosophical concepts of ultimate explanations (or attempts to undermine them) only incidentally, more for the sake of a fuller picture of the philosophical ideas involved than of their analysis in depth. Secondly, out of all the versions and interpretations of the creation concept I shall only select ones which may be referred in one way or another to contemporary science, or those which, albeit historically distant from the present times, are still indispensable for the right understanding of such reference. This criterion is not so restrictive, as the history of ideas, in science as well as in philosophy, shows that ever since Christian Antiquity right through to the modern period, the mainstream thought on creation has been strictly linked genetically with the evolution of the ideas which led up to the development of the modern sciences. However, it is not my intention to compile a history of those genetic links, but to try to look at ultimate explanations from a different perspective than the one usually taken up on the grounds of physics and cosmology.

Is the philosophical and theological speculation located on a long extension of the investigations based on scientific theories and models? Perhaps the two are somehow mutually complementary to each other? Or perhaps – as some people claim – although apparently concerned with much the same thing, scientific inquiry and theological and philosophical deliberation are mutually untranslatable? Regardless of which of these possibilities (or maybe yet another one) is true, all of them are an expression of the same instinct ingrained in human rationality: to leave no stone unturned in seeking for an answer to every valid question.

PART I

MODELS

Chapter 2

⌒

PROBLEMS WITH THE ETERNITY

OF THE UNIVERSE

1. THE ETERNITY AND INFINITY OF THE UNIVERSE

*O*ne of the simplest ways to explain the world is the attempt to convince oneself that there is nothing to explain. If the universe has always existed, then there is nothing to explain. Reality is simply "given us" and the problem is removed. No wonder that the doctrine of the "eternity of matter" has always constituted one of the pivotal claims of all manner of materialisms.

But such an explanation is only apparent. Already St. Augustine observed that if someone were to stand barefoot on the beach for all eternity, then his footmark on the sand would be eternal too, but nonetheless it would still have its cause – the foot making it. If we wanted to neutralise this argument as well, we could query the sense of asking about any kind of cause. This device was employed in the diverse forms of Positivism: it was claimed that experience can inform us only of the sequence in which phenomena occur, but not of their inner causal relations. This type of therapeutic manoeuvre has survived only within some of the more exotic trends in philosophy. Various sciences relating to the world are still searching for causal chains within those aspects of the world subject to their fields of study.

It is a historical fact that for a long time, more or less from the French Enlightenment onwards, the belief in the "world's eternity" has generally been regarded as something in the way of an ultimate explanation with no further questions asked relating to other "deeper causes of existence." Admittedly, the image of an eternal world has been consolidated by the progress made in classical physics. Newton himself was deeply convinced that his mechanics, when applied to the system of

15

M. Heller, *Ultimate Explanations of the Universe*, DOI 10.1007/978-3-642-02103-9_2,
© Springer-Verlag Berlin Heidelberg 2009

the universe, called for a Grand Architect to fix the initial conditions for the laws of mechanics, but his concept of absolute space and absolute time set the stage on which processes could take place without being influenced by space and time. True enough, the differential equations describing the laws of nature require initial conditions, but these may be selected for any arbitrary moment in time. Thus the word "initial" turns out to be established purely by consensus, and the initial (or boundary) conditions themselves serve only to enable us to select the right solution out of the entire class of possible solutions, therefore they do not appear to give rise to any serious problems from the philosophical point of view.

In the eighteenth and nineteenth centuries the image of an eternal universe extending out to infinity in a Euclidian space was one of those beliefs which are so obvious that they are not even discussed. This did not mean, however, that there were no problems pertaining to this image – generally accepted beliefs need not be unquestionable. Newton himself observed that his law of gravitation applied to an infinite universe containing an approximately homogeneous distribution of stars generated serious problems. How could the stability of such a system be ensured? An arbitrarily small disturbance in the density of the distribution of the stars would cause the collapse of the entire system into one gigantic body.[1] In 1895 the German astronomer Hugo von Seeliger said that this problem was so fundamental (and today it is called Seeliger's paradox) that he put forward an alternative. We should either query the infinity of the universe, or amend Newton's law of gravitation. And he decided on the latter option. A year later and absolutely independently, Carl Neumann, a mathematician, did the same. Both Seeliger and Neumann proposed that an additional constant be introduced to Newton's laws to stabilise the system of the universe.

2. THE THERMAL DEATH HYPOTHESIS

The belief in an eternal universe was well-nigh a dogma of the mechanistic worldview. The emergence of thermodynamics in the nineteenth century was hailed as yet another success for this philosophy. The theory of heat based on the concept of phlogiston, a "thermal fluid" flowing from warmer bodies to colder ones, was successfully replaced by statistical mechanics, in other words simply the Newtonian mechanics, in which the mean values of various magnitudes referred to large numbers of material molecules. However, this success cast a shadow over the doctrine of the eternal universe.

The first law of thermodynamics is the law of the conservation of energy applied to heat changes. So far there are no problems looming ahead. If we consider the universe as a single large thermodynamic system, the first law of thermodynamics may be regarded as an argument in favour of an eternal universe. If the energy of such a system is conserved, then it must have always been so, since energy can neither be lost not created.

The second law of thermodynamics was formulated in 1850 by Rudolf Clausius, who expressed it in the form of a theorem that no machine can be constructed which can transfer heat from a body at a lower temperature to a body at a higher temperature. Four years later he gave this principle a more mathematical form, introducing a function which he later named entropy. The principle expresses the tendency prevailing in an isolated thermodynamic system to equalise temperature, and it takes the form of a theorem which says that in such systems entropy increases (or remains constant in systems with reversible processes). Clausius himself did not refrain from drawing cosmological conclusions, observing that the entropy in the world was tending to a maximum, that is to the establishment of a uniform temperature throughout. Later Hermann Helmholtz called this state the heat death of the universe.

William Thomson drew further conclusions from the second law of thermodynamics. If heat death has not occurred yet, then the cooling down of the universe (viz. the equalising of temperatures to a uniform value throughout) must have started a finite time ago, in other words the process must have had a beginning. Thomson wrote of "some finite epoch [with] a state of matter derivable from no antecedent by natural laws."[2] This was too reminiscent of the notion of a beginning of the world not to evoke controversy and heated debate. They still recur even today in diverse publications.

Nonetheless most scientists did not treat all these cosmological discussions and speculations very seriously. The well-known Irish astronomer Agnes Mary Clerke expressed the prevailing opinion when she wrote in 1890 that whatever lay beyond the boundaries of the Milky Way was not the subject of scientific study, since "with the infinite possibilities beyond, science has no concern."[3] Cosmology would not become a respectable science until Albert Einstein and his general theory of relativity. The consolidation of the relativistic cosmology was a process which went on for several decades in the twentieth century, starting in 1917 with Einstein's first paper on cosmology.

3. EINSTEIN'S FIRST MODEL

Already at first glance Einstein's article is extremely pioneering, though otherwise it looks just like a standard scientific paper.[4] When he wrote it Einstein had the field equations available for the general theory of relativity, which show the gravitational field as a deformation of space-time caused by the distribution of all the sources of the field. In such a situation the cosmological question appears quite naturally. We simply have to answer the question in what way the mean distribution of the sources of the gravitational field deforms the space-time geometry. Of course in answering that question we have to resolve a whole series of conceptual and technical issues. The way in which Einstein accomplished this made his paper a breakthrough.

Above all, since Einstein's field equations are a set of differential equations there is the problem of boundary conditions, which is in turn connected with the distribution of the sources of the gravitational field (Einstein simply spoke of stars). The natural solution often applied in astronomical enquiries is to assume that we are dealing with an isolated system of bodies, on which the gravitational influence of other bodies is negligible enough to be ignored (the gravitational field disappears at the "boundary of the problem"). In cosmology this would correspond to one "island of stars" (e.g. the Milky Way) in the empty space surrounding it. Astronomers had been debating for quite a long time over "the island distribution of matter": some held that the spiral nebulae visible with a telescope were clouds of dust and gas in our own galaxy (the Milky Way), while others said that they were different galaxies, separate "island universes". The dispute continued, but for the time being neither side could put forward a clinching argument. Einstein probably did not know of the controversy, but in a sense he resolved it with one sweep of the pen. Considering the issue of boundary conditions, he observed that the assumption that the gravitational field disappeared at infinity could not hold in cosmology. A simple statistical approach convinced him that if we assumed just one "island of stars" in an otherwise empty space, then sooner or later the stars, agitated by random motion, would have to evaporate from the island. A solitary galaxy would be an unstable structure. Therefore we have to assume a statistically uniform distribution of stars (galaxies or clusters of galaxies, in the terminology used today) in space.

But then what boundary conditions should be applied? Somewhat earlier the Dutch astronomer Wilhelm de Sitter had hinted at a solution. In Einstein's theory we do not have to insist on a flat space: we know that gravitation distorts its geometry. So we may do away completely with the "boundary," and hence with the need to adopt any kind of boundary conditions. Such a situation would hold if space were spherical in shape, analogous to the surface of a sphere (if we move along it, nowhere do we encounter an edge). Einstein calculated that there was a solution to the field equations which had these properties.

There was just one remaining problem, the one that had troubled Newton – the question of gravitational instability: why would the stars in a spherical universe not collapse into a single point? To obviate the difficulty, Einstein did what von Seeliger and Neumann had proposed earlier with respect to Newton's theory of gravitation: he augmented his equations with a component entailing a constant the purpose of which was to stabilise the model. This constant – Einstein named it the cosmological constant – is an exact counterbalance of the attracting gravitational force. That is how the first cosmological model based on the theory of relativity was constructed. Today we call it Einstein's static model.

4. THE UNIVERSE AND PHILOSOPHY

Let's not be led astray by appearances, however. True enough, Einstein's paper is an example of a fine piece of research opening up new horizons while at the same time addressing the old problems. But his aim went much further: it was precisely to reach the ultimate explanation. Naturally, such intentions are not to be disclosed in a research paper submitted for publication in a scientific journal, although they may often inspire many an author. On the other hand we have to admit that Einstein cared far less about conventions than many of his colleagues. The attentive reader will quite readily identify a certain philosophical motif in his 1917 paper: "In a consistent theory of relativity," he wrote, "there can be no inertia relatively to 'space,' but only an inertia of masses relatively to one another."[5] Again this sounds technical, but it's fairly easy to decipher what Einstein was thinking of. The inertia of a particular body with respect to space, which would have to be something like Newton's absolute space, would mean that the body's mass, which is a measure of its inertia, would be its absolute property, something with which the body was endowed a priori. But the world should be a "closed system," all of its justifications should remain within it, not assumed a priori. The only sensible solution to this situation was to assume that the mass of a particular body was as it were induced in it by all the other bodies in the universe. Hence there would be no inertia with respect to space, only with respect to other masses. Einstein took this idea from the writings of the physicist and philosopher Ernst Mach, and in his honour called it Mach's principle. The intention to create a theory of physics incorporating Mach's principle was one of the main motives behind Einstein's efforts which eventually led to the emergence of the general theory of relativity. No wonder that this motive is clearly visible in his first paper on cosmology.

But Einstein's philosophical inspirations went even further. Ever since his young days he was interested in the life and work of Baruch Spinoza, a seventeenth-century philosopher. Spinoza was so fascinated with instances of rationality in the world that he identified the world with God. "By God," he wrote, "I understand a being absolutely infinite, that is, a substance consisting of an infinity of attributes, of which each one expresses an eternal and infinite essence."[6] Understood in this way, God is identical with the universe; hence God is the "substance" which exists "of itself" and is "self-explanatory." Einstein was quite open about his sympathy with pantheistic views of this kind. He, too, was fascinated by the "rationality of the universe" and often spoke of his "cosmic religion" in connection with this. No wonder, then, that the universe was "to explain itself"; the right cosmological theory should be the ultimate theory.[7]

Einstein immediately took up de Sitter's suggestion that troublesome boundary conditions could be evaded by assuming that the universe was spatially closed. The logical enclosure of the universe, that is the idea that all of its explanations should be enclosed within the universe, found its expression in the geometrical enclosure of the universe. On finishing his paper Einstein had every reason to feel pleased with himself. There was only one solution to the gravitational field equations which met all of his philosophical criteria. That solution presented an eternal universe, spatially closed and obeying Mach's principle.

Einstein thought that the "cosmological problem" had been solved. I wonder what research problems he was pondering about after that?

5. AN EXPANDING VACUUM

The "universe's rationality" is indeed one of its fascinating features. It certainly needn't have been so that our minds would be capable of fathoming the mysteries of its structure. For we have managed to fathom so much. Einstein's first paper on cosmology was undoubtedly a milestone on the road to understanding cosmic structure. As we think about this a disconcerting question comes to mind: are our brains advanced enough to allow us to completely solve the mystery of the universe? Or to put it in another way; does the structure of the universe have to correspond to our brain structure to such an extent as to allow us full access to discovering the way it works? On finishing his paper Einstein did not realise how far he still was from ultimate solutions. But he was soon to find out.

Still in 1917 de Sitter published a paper presenting a new cosmological solution to Einstein's equations (with the cosmological constant).[8] In this paper de Sitter embarked on a dispute with Einstein's understanding of Mach's principle and put forward his own interpretation. But this was not what proved fatal to the views of Einstein. The very existence of de Sitter's solution put them to a difficult test. In de Sitter's solution the density of matter is equal to zero. In other words de Sitter's model is empty, and in spite of this the structure of space-time is still well-defined. Therefore it is not defined by means of a distribution of "material sources" and Mach's principle (as Einstein understood it) is not obeyed in the general theory of relativity. Soon it turned out, thanks to the work of Georges Lemaître,[9] that although de Sitter's world was empty, his space was expanding: if we were to put into this world two particles the masses of which could be ignored as negligible, so as to still be able to consider the model as empty, then those masses would begin to move away from each other.

Meanwhile ever since 1912 Vesto Slipher had been measuring shifts in the spectra of galactic nebulae. In 1918, on the basis of his own and Slipher's observations, Carl Wirtz expressed an opinion that the prevalence of red shifts in the spectra of nebulae could mean that these nebulae were moving away from each other. In the same year Eddington wrote in a letter to Shapley that the spreading out of the nebulae had been predicted in de Sitter's model.[10] The recession of the nebulae came to be known as the de Sitter effect.

6. THE CRISIS OF EINSTEIN'S PHILOSOPHY

From the theoretical point of view the situation was paradoxical. Einstein's model had a non-zero density of matter but did not predict the moving away of the galaxies (spiral nebulae). De Sitter's model predicted the moving away of the galaxies but had a zero density of matter. Nonetheless the argument that the mean density of matter in the real world was smaller than the best vacuum we could obtain in laboratories on Earth, in other words that we could treat de Sitter's model as a close approximation to reality, was a dodge. And scientists knew it. After all, theoretical zero density is not the same as a very small density.

But the paradox was soon resolved. The Russian mathematician and meteorologist Aleksandr Aleksandrovich Friedman published two papers presenting his discovery of a whole class of spatially homogeneous and isotropic solutions to Einstein's equations of which Einstein's and de Sitter's solutions were special cases.[11] In this class there was only one static model (Einstein's); all the others were either expanding or shrinking. He also explained the apparently paradoxical status of de Sitter's solution: all the models expanding out to infinity (monotonically) tended to de Sitter's empty model as time tended to plus infinity. Thus de Sitter's state was effectively an asymptotic state for the expanding models, in which the density of matter tended to zero in outcome of the expansion.

Gradually the situation was starting to clear up. Einstein's proposition that there was only one unique, uniquely possible cosmological model concordant with all the philosophical expectations turned out not to hold. In cosmology, as in all the other branches of physics, many models can be constructed and only observation will tell which of them corresponds best to the reality in the world.

Cosmology would not become a fully experimental science until the last decades of the twentieth century, but it started to mature already by the 1920s. In 1929 Edwin Hubble collated about 40 results for the red shift measurements in the spectra of galaxies and published his famous law: the velocity at which a galaxy is moving away is directly proportional to its distance from us.[12] These results were already in circulation among scientists. In 1927 Georges Lemaître compared the results of measurements of the red shift with predictions for one of the solutions discovered by Friedman, which he had found independently of Friedman, and confirmed that there was no discrepancy between the theory and observations.[13]

In the 1930s the paradigm of an expanding universe became firmly established. Even Einstein, who for a long time would not accept it, finally had to concede in the face of facts. The reason for his opposition had been that an expanding universe suggested the idea that it must have had a beginning. Knowing the distance to a few galaxies and the velocities at which they were receding, on the basis of Hubble's law it is easy to estimate how long ago all the galaxies were situated "at one point." For Einstein this was a difficult conclusion to accept. A universe which was supposed to be self-explanatory should not have a beginning.

Chapter 3

◠

A CYCLICAL UNIVERSE

1. THE PROBLEM OF THE BEGINNING

*B*y about the mid-1920s it was evident that an eternal universe could not be kept on in relativistic cosmology "at a low cost." Not only does Einstein's static model contradict observations and experimentally measured values for the red shift in galactic spectra, but – as Eddington showed – it is also unstable: it cannot persist in a state of "static equilibrium," and the occurrence of even a slight perturbation, which is what gravitation does by its very nature, would give rise either to its collapse or expansion. Thus all the indications are that the universe is not static but dynamic. But a dynamic universe implies the question of a beginning. The first measurements of the red shift suggested, and Hubble's subsequent work confirmed, that the universe is expanding; and if it is expanding, then by extrapolating back in time we reach a conclusion that the process must have started from a state in which all the matter and energy in the universe now were in a state of gigantic compression. The name "Big Bang" had not come into use yet, but the idea itself was crystallising out and becoming well-established.

As he once admitted in a discussion with Lemaître,[1] Einstein did not like the concept of an expanding universe because it conjured up a conclusion "too reminiscent of the idea that the universe had been created." Admittedly, in the early years of the development of relativistic cosmology notions derived from philosophy and researchers' worldviews played a substantial part. But it is also true that in scientific cosmology the idea of a beginning is unwelcome, not the least for purely methodological considerations. In classical physics, to which the theory of relativity and cosmology certainly belong, whenever we have any kind of

M. Heller, *Ultimate Explanations of the Universe*, DOI 10.1007/978-3-642-02103-9_3,
© Springer-Verlag Berlin Heidelberg 2009

evolutionary process we explain it by means of dynamic equations. The following rule applies to such equations: if we know the state the system is in and its rate of change at a given time (or its state at two different moments in time), we are able to calculate its state at any arbitrary moment. This is known as the principle of determinism. In accordance with this principle, the dynamic equations for a given system and its rate of change at any given instant in time (or its state at two instants in time) give a full explanation of the particular evolutionary process. The equations Friedman used in his 1922 and 1924 papers to describe the evolution of cosmological models are dynamical systems in the sense used in classical physics, but in spite of this they fail to give a classical (deterministic) explanation of the evolution of the universe: knowing the state of the universe and its rate of change at any arbitrary instant in time (or its state at any two instants in time), it is not possible to calculate the states of the universe "prior to the beginning" using these equations. The classical principle of determinism breaks down at the beginning (and also at "the end," for models which envisage an "end").

In Friedman's models the "beginning" and, by analogy, the "end," may be described as follows: if time tends to the "beginning" (or "end"), then the mean density of matter tends to infinity. In more detailed models the same holds for certain other physical parameters, e.g. temperature and pressure. But in physics states in which some values tend to infinity are considered "non-physical," since "you cannot do anything" with such values. In particular they defy all attempts to measure them, even estimates carried out on a purely theoretical basis. Not only does "the beginning" violate the classical deterministic explanation, but it also brings non-physical components into the model. For this reason the term *singularity* (*initial* or *final*) was introduced for the "beginning" and "end" in cosmological models. As we shall see in due course, the occurrence of singularities in models of the universe became a serious problem in cosmology.

2. AN OSCILLATING UNIVERSE

In view of the fiasco of Einstein's static model, which was to represent a universe that had always been in existence, impelled by scientific and/or philosophical considerations, cosmologists and other thinkers turned their attention to oscillating relativistic models. The idea of a pulsating world passing through an endless series of "beginnings" and "ends," had been present in the history of human thought for a long time. It was also an embodiment, albeit in a different manner, of the concept of an eternal world, in other words a self-explanatory world (again in a certain sense).

The solutions Friedman discovered in his 1922 and 1924 publications comprise an infinite number of solutions, among them also ones representing oscillating worlds. In the class of models with a constant, positive curvature of space (closed models) oscillating solutions exist for values of the cosmological constant Λ less than Λ_E, the cosmological constant for Einstein's static model. In particular, there exists an oscillating solution for $\Lambda = 0$. Oscillating solutions also exist for negative values of the cosmological constant.[2] In the class of models with a zero or a negative curvature of space (open models) oscillating solutions exist for negative values of the cosmological constant.[3] In both classes, for open and closed models, the smaller the value of the cosmological constant, the shorter the period of oscillation.

To describe a cosmological model's evolution let us introduce a function $R(t)$ called the *scale factor*. We may envisage it as the mean intergalactic distance. The scale factor is a function of time t. For the static model it is a constant function, $R(t) = $ const. If the model is expanding, $R(t)$ is a function which increases with time; if the model is shrinking, $R(t)$ is a function decreasing with time. The typical path of evolution for an oscillating model is as follows. The cycle begins with the initial singularity, for which $R(t) = 0$. Then comes the expansion phase, during which all the galaxies recede from each other until $R(t)$ reaches its maximum; from that moment $R(t)$ begins to decrease and the expansion passes into contraction, the galaxies move closer and closer to one another, until finally, when $R(t)$ again goes to the zero value they collapse to the final singularity.

Strictly speaking, at both the initial and final singularity the solution to the dynamic equations breaks down. There is no known method of extending the solution beyond the singularities. Essentially what we have is not an infinite number of cycles, but just one: not an oscillating or pulsating model, but just one pulse. The solution gives us no information on what was there before the initial singularity, or what will happen after the final singularity. Since we have no knowledge on this subject, we may imagine whatever we like. On these grounds many cosmologists have imagined an infinite number of oscillating cycles, making the reservation that although we do not know the mechanisms for "rebound," viz. the passage from one cycle to the next, it would be reasonable to assume that before an expansion there was a contraction phase. Often a remark would be added that at the turning point the scale factor $R(t)$ was not quite zero, but had a very small value, not actually equal to zero. In this way there would be no singularity at the beginning of each cycle, only a state "of very high density." Such assumptions would be supported merely by the expectation that the future development of the theory would confirm them.

CHAPTER 3

3. THE RECURRENCE THEOREM

The question of "eternal returns" appeared in science independently of cosmology thanks to Poincaré's famous theorem, known as the recurrence theorem, which this scientist proved in 1890 in his paper on the three-body problem.[4] As formulated by Poincaré, the theorem says that in a system of material points subject to forces which depend only on position in space, the state of motion, determined by configuration and velocity, after a certain time will return, with an arbitrary approximation, to its initial state once more or even an infinite number of times, providing the coordinates and velocities do not increase to infinity.

The natural application of Poincaré's theorem was classical mechanics. The theorem holds for a finite mechanical system[5] in the phase space on which there is defined a finite measure needed to determine the evolution of the system.[6] Poincaré was aware of the possibility of exceptions occurring in which the system will return to its initial state only a finite number of times or even no times at all. This issue was clarified by Constantin Carathéodory, who showed that the measure zero should be assigned for the set of exceptions.[7]

The conclusions to be drawn from Poincaré's theorem were too much of a temptation not to be applied to speculations on the history of the universe. However, the idea of a universe that every so often returned to its "starting point" appeared to contradict the second law of thermodynamics which, if extrapolated to the cosmic scale, suggested a one-way cosmic history, running from a state of minimum entropy to a state of thermodynamic equilibrium characterised by maximum entropy and referred to as heat death. Already William Thomson had reached such conclusions on the grounds of the second law of thermodynamics, writing of "some finite epoch [with] a state of matter derivable from no antecedent by natural laws."[8] The hypothesis of heat death had appeared as such in the work of Hermann Helmholtz.[9] Later speculations of this kind gave rise to a long series of discussions and debates[10] in the course of which Ernst Zermelo observed that there was a certain inconsistency between Poincaré's theorem and the heat death hypothesis: a cyclic history of the universe and a one-way process towards heat death cannot both be true simultaneously.

The problem was clarified by Ludwig Boltzmann, who showed that the laws of thermodynamics were statistical in character, what is more only over a long time scale. Even if the universe were to reach a state of thermodynamic death, statistical fluctuations could displace it from that state. Furthermore, according to Boltzmann, we may conceive of the universe as having reached a state of heat death long since, with only "our world" as a "small fluctuation" in it with a rising entropy.[11]

26

Although originally Poincaré's theorem was applied to classical mechanics, after the emergence of relativistic cosmology it seemed natural to apply the theorem to spatially closed models of the universe. The spatial closedness should be regarded as a counterpart of the concept of an isolated system in thermodynamics, and Friedman's oscillating models provide the relativistic version of the cyclic nature of the universe. We shall return to this issue in due course.

4. TOLMAN'S UNIVERSES

Oscillating models of the universe and thermodynamic problems in the context of the general theory of relativity pretty soon attracted the attention of the American physicist Richard Tolman. Already in 1934 he published a monograph entitled *Relativity, Thermodynamics and Cosmology*,[1] which is sometimes still referred to even today, not only for historical reasons. In this book he conducted a detailed analysis of the laws of thermodynamics from the point of view of their agreement with the special and general theory of relativity.

Tolman realised that in relativistic cosmology it was an overstatement to speak of oscillating models of the universe. Every phase of contraction would have to end in a singularity at which the solution to Einstein's equations broke down and, strictly speaking, it was no longer possible to predict what would happen afterwards. However, he thought that this was a weakness of the contemporary theory rather than a fundamental obstacle. Thus he made the working assumption that when the universe reached its "minimum volume" some sort of hitherto unknown physical mechanisms would emerge to "make the universe rebound" and initiate the expansion phase. On the basis of this assumption Tolman proceeded to examine sequences of cosmic oscillations unlimited by time.

We should distinguish between reversible and irreversible oscillations in these sequences. In the cosmological models studied hitherto it had been assumed (usually tacitly) that entropy remained constant, viz. there were no processes taking place in them involving the dissipation of energy. But for the analysis of the thermodynamics of oscillation we should introduce energy dissipation. In spatially isotropic models, viz. ones in which the expansion is uniform in all directions, neighbouring layers of cosmic matter "do not rub against each other" and therefore there is no dissipation of energy. However, Tolman assumed that energy could be dissipated at the expense of the potential energy stored in the gravitational field, although he did not name any specific physical mechanisms which might carry out this dissipation. It was not until much later that researchers realised that these mechanisms were associated with the so-called bulk viscosity, which causes the dissipation of energy in outcome of a rapid change of volume.[13] On the basis of his assumption, Tolman showed that if irreversible processes (associated with the dissipation of energy) were taken into consideration, then the amplitude of successive cycles in the oscillating model would increase. This was something of a surprise, since in classical mechanics if energy dissipation is envisaged in an oscillating system (an oscillator), the oscillations diminish and die down. But according to Tolman's calculations the reverse should happen in the general theory of relativity – the amplitude of the oscillations should increase. Tolman was right – this happened because the system could draw an unlimited amount of energy from the gravitational field (viz. the curvature of space-time). Years later, when terms responsible for bulk viscosity were brought into the cosmological equations, in accordance with the strict rules of the game, it turned out that Tolman had missed another effect: not only does amplitude increase for successive cycles, but also the cycles become asymmetric with time; contraction takes place faster than expansion. Time asymmetry means that the processes associated with energy dissipation determine the arrow of time. It also turned out that Tolman's type of solutions included ones for which the oscillations could not be extrapolated back in time to infinity (quite apart from the singularity problem): the earlier an oscillation, the shorter its period, until it was eventually "reduced to zero." Here, too, loomed a vision of a beginning.[14]

5. TIPLER'S THEOREM

If in the general theory of relativity some dynamic questions may take a form so drastically different from their counterparts in classical mechanics, we should ask whether in the Einsteinian theory there is a counterpart of Poincaré's theorem of recurrence, and if so, then what does it say. The answer to this question was found by Frank Tipler, who in 1980 proved the existence of a relativistic counterpart of Poincaré's theorem.[15] Tipler's theorem is expressed in highly technical language, and the proof calls for advanced mathematical tools; below I present just the basic idea of the theorem.

We want to learn whether the relativistic universe will one day return to a former state. By the state of the universe at any moment in time t we mean the set of all the events taking place in the universe at time t. In the technical language used by cosmologists this is called a space *section of the universe at time t* (or an instantaneous *section of the universe*). If the initial conditions determining the further development of the universe are determined on such an instantaneous section, it is called a *global Cauchy surface*. A cosmological model is *time-periodic* if it has two identical global Cauchy surfaces with the same initial conditions at two different moments of time.[16] Two Cauchy surfaces of this kind represent the same state of the universe. Thus a time-periodic model describes the universe returning to a former state. By analogy[17] we may describe the return of such a universe to a state close to a former state.

Tipler's theorem states that if space-time *M*

1. contains a closed[18] Cauchy surface such that the initial data for it determine the entire history of the universe;
2. gravitation is an attractive force;
3. and every history of a particle or photon experiences an attractive gravitational force at least once – then space-time *M* cannot be time-periodic.

In other words, if the conditions of this theorem are fulfilled, then the universe *cannot* return to a former state. The last two conditions are very tolerant and we should expect them to be met in the real universe. The condition of spatial closedness is essential, for if it is removed the theorem cannot be proved. Moreover, examples are known of spatially open worlds which are time-periodic, although all the other conditions of Tipler's theorem are observed in them. One such model is an empty world, appropriately symmetrical, with just one, static star. Admittedly, it is not very realistic as a practical proposition for the description of our universe. However, it falsifies Tipler's theorem for open models.

It is noteworthy that classical determinism was one of the salient assumptions in Poincaré's theory of recurrence, whereas in the general theory of relativity this condition (1. in Tipler's theorem) is one of the factors leading to the conclusion ruling out eternal returns.

On the basis of this result and other reflections in this chapter we may reach a conclusion that ideas intuitively drawn from classical physics should not be transferred uncritically to relativistic physics. On its largest scale the universe is relativistic, and hence global cosmological conclusions should be reached on the grounds of precise analysis rather than in a flash of intuition.

6. SINGULARITIES

So does the concept of eternal return have a chance of fulfilling the function of an "ultimate solution" in contemporary cosmology? As we have seen, the fairly appealing, commonsensical idea that the history of the universe is made up of an infinite series of cycles comes up against a number of serious obstacles. We cannot say that the cosmological model corresponding to this idea has been abandoned altogether, but at present it is undoubtedly creating more conceptual problems than it is resolving.

Currently the most serious difficulty in this model seems to be the occurrence of singularities at the beginning and end of each cycle. Tolman and his contemporaries might have entertained the hope that the singularities were an artefact effected by the adoption of oversimplified assumptions for the construction of the model. Very often the assumption of a homogeneous and isotropic nature of the universe was suspected as responsible for this. Both Einstein and Tolman expressed such an opinion. However, already Lemaître's early research[19] had shown that singularities still occurred in the cosmological model even if the assumption of isotropy was cast aside. On the contrary, the removal of this assumption increased the tendency of singularities to occur. In the 1960s Stephen Hawking, Roger Penrose, Robert Geroch and others[20] proved a series of theorems indicating that the occurrence of singularities in space-time theories like the general theory of relativity was the rule rather than the exception. Moreover, the initial and final singularities in the Friedman-Lemaître models belong to the class of strong curvature singularities and are characterised by a breakdown of the structure of space-time (in other words the concept of space-time becomes meaningless in them); and hence we may speak of only one cycle in them for the history of the universe, which starts with the initial singularity and ends with the final singularity. No solution can be prolonged beyond the singularities.

It should be stressed, however, that theorems of the occurrence of singularities apply only to "classical singularities," that is analyses which do not take the quantum effects of gravity into account. This offers an escape route for the avoidance of such theorems. Perhaps the quantum effects of gravity will breach one of the conditions in the theorems for the occurrence of singularities, thereby facilitating a smooth transition from the contraction phase to the expansion phase. Many scientists have set their sights on this possibility, which looks appealing from the vantage-point of the interests of the search for ultimate explanations. However, the snag is that hitherto we have not yet worked out a generally acknowledged and experimentally confirmed quantum theory of gravitation, and the diverse trends in the research and the partial results obtained in the most popular approaches such as the theory of superstrings or Ashtekar's loop, have not yielded an unambiguous answer in this respect. Nonetheless we may observe a distinct trend: authors have a clear preference for solutions in which there are no singularities at all, or else the singularities seem easy to remove. What is more, they tend to treat precisely these attributes of their model as the criteria making it appealing.

To conclude this chapter I would like to relate a certain episode from the history of science which should serve as a warning to all those who are guided in their choices in science by grounds other than mathematical consistency and experimental verification. In the nineteenth century, when the heat death hypothesis cast a shadow of doubt over the concept of an eternal universe, W.J.M. Rankine[21] put forward a conjecture that the energy dissipated in the universe (on the grounds of the second law of thermodynamics) would one day come up against a barrier in "the interstellar ether" situated at a finite distance away from the Earth, rebound from it, and once again accumulate in diverse "foci." This process was supposed to be periodic, which would ensure the world of eternal existence. You can do it that way if you like, but reasoning on the strength of this strategy gives ultimate explanations which fade away into oblivion within a few years.

Chapter 4

⁓

A LOOPED COSMOS

1. VISIONS OF CLOSED TIME

O ne of the more macabre ideas of how to eliminate the beginning from the history of the universe is the concept of closed time: a sequence of events recurring an infinite number of times; a history heading nowhere but only endlessly reiterating what has already occurred an infinite number of times; an unending chain of births, deaths, and renewed births; the hopelessness of the impossibility of wresting free from an inexorable loop. Nonetheless this idea has been resurging quite often, both in our none too strictly controlled imagination as well as in the history of human ideas.

In Antiquity the Stoics reduced the idea of eternal returns to its logical extreme. The history of the world was cyclical: after each cycle the world returned to its original state (*apokatastasis*), passing through a phase of destruction by fire and then starting a process of ordering itself anew (*diakosmesis*). In each cycle exactly the same structure was reconstructed, down to its most minute detail: "After the passage of centuries the same Socrates will be teaching in the same Athens, and in the streets of the same cities the same people will be going through the same suffering."[1]

Every so often this doctrine is self-renewed and arises out of its own ashes. A contribution to its popularity in the modern period has come from Friedrich Nietzsche, who was very fond of it and treated it as a sort of religious message. He also tried to find a scholarly justification for it, albeit rather ineptly. In his opinion, the world should be envisaged as "a particular number of foci of force" and therefore "had to go through a calculable number of combinations, as if in a game of dice, in the grand game of existence." Hence the world was a

M. Heller, *Ultimate Explanations of the Universe*, DOI 10.1007/978-3-642-02103-9_4,
© Springer-Verlag Berlin Heidelberg 2009

sequence of identical combinations "which had already been repeated an infinite number of times and which continued to play out their game *ad infinitum*."[2]

You might think that today the idea of looped time could persist only in literary visions and science fiction. But the history of science turns out to be stranger than fiction.

2. KURT GÖDEL'S UNIVERSE

Quite out of the blue it turned out that the general theory of relativity lends fairly strong support to the concept of closed time. The first solution of Einstein's equations involving closed time-like curves was discovered in 1924 by Cornelius Lanczos, who was later Einstein's assistant.[3] It was rediscovered in 1937 by Willem Jacob van Stockum, a Dutchman who was killed in action during the Second World War, as a pilot fighting for the Allies.[4] This solution (now known as van Stockum's dust) describes a space-time with a cylindrical symmetry, in which matter in the form of dust rotates around an axis of symmetry. This fact physically distinguishes the axis of symmetry, as a result of which the space-time is not isotropic. Van Stockum's solution has one other feature, apart from closed time-like curves, for which it is hard to give a physical interpretation: the density of the dust particles increases with distance away from the axis of rotation.

In spite of their exotic properties, neither Lanczos' nor van Stockum's solution attracted much notice. It was not until Kurt Gödel's discovery of another solution in 1949 that people's attention was turned,[5] most probably thanks to the fact that Gödel was already a well-known personality and also because from the very start he promoted his solution as a cosmological model. Gödel's solution entailed closed time-like curves, and understandably the possibility of a return to one's own past stirred up a sensation. To reach a closed time-like curve in Gödel's world you would have to have an unrealistically immense store of energy available to accelerate your spaceship appropriately, but what was that in view of the prospect of conquering time? Let's take a closer look at Gödel's solution.[6]

Gödel's universe is filled with matter consisting of dust with a constant density, just like the Friedman-Lemaître standard models. His space-time is flat and homogeneous (it is called R^4 space) and has rotational symmetry around an axis. This axis may be identified as the trajectory of a particle with an initial velocity in the radial direction. Hence it may be said that the matter in Gödel's model is rotating around this axis, or – equivalently – the axis is rotating with respect to the matter at rest. But this time the "axis of rotation" is not distinguished in any way at all. It may be transferred to any arbitrary point by a simple change of coordinates, such that the history of any arbitrarily chosen particle moving along a radial path may serve as Gödel's axis.

The rotational symmetry of Gödel's world is associated with a phenomenon which has already been mentioned – a closed time-like curve passes through every point in Gödel's universe.[7] In other words in Gödel's solution there is no cosmic time capable of "increasing at a uniform rate" with the history of any observer or particle (of non-zero rest mass).

There are no singularities – neither an initial nor a final singularity – in Gödel's model,[8] and hence the "haunting prospect of a beginning" has been eliminated from it. But such a model is just a purely mathematical option, since it does not incorporate the effect of an expanding universe, in other words this model offers no explanation for the red shift observed in galactic spectra – a phenomenon which quite definitely exists in the real world.

There are many indications that Gödel started his search for a solution prompted by his own philosophy of time. Shortly after publishing his model he wrote a separate paper presenting his views on this subject.[9] He believed that time could be objective (real) only if there existed an infinite number of successive "layers of the present," one following another. But the special theory of relativity rules out such a possibility. The situation seemed to be saved by the fact that in all the cosmological solutions to Einstein's equations known at the time there exists a global time, which makes a succession of "layers" of the present possible. However, the solution discovered by Gödel shows that it is not a typical situation, but enforced by a symmetrical distribution of matter, thanks to which it was possible to apply a privileged system of coordinates extending over the whole of space-time in such a manner that one of the coordinates may be interpreted as global time. In the general case there was no such thing as global time nor an absolute "present moment." Hence – according to Gödel – time may not be considered objective; it was only a figment of our imagination projected onto the universe.

However, the great logician seems to have made a mistake: he treated the absoluteness of time (and simultaneity) as identical with its objectivity. But the relative need not be subjective. Dependence on a system of reference may be, and often is, an objective fact. We are not obliged to agree with Gödel's intuitions, but we should appreciate his difficulties. The theory of relativity had introduced a host of new concepts and a lot of time had to pass before physicists, philosophers and others who were interested in the issue could cope with all this and take stock of it intellectually.

There was an "existential" backdrop to Gödel's grapple with time. After his death the notes he left revealed that for nearly two decades he had been searching for a theoretical possibility to overcome death by making use of closed time-like curves.[10]

Gödel's solution triggered an avalanche of papers in which more and more solutions to Einstein's equations were found not only containing closed time-like curves but also exhibiting numerous temporal and causal pathologies.[11]

Although Gödel's solution does not offer a description of the real world, it has played an important role in the development of the mathematical methods applied in the general theory of relativity. From the very beginning physicists and philosophers, as well as many bystanders, have been intrigued by the problem of time in Einstein's theory. Many believed that the time riddles would be "straightened out," or at least grasped, if the theory were presented in the form of an axiomatic system. A few attempts were made to axiomatise it, undertaken by scientists like A.A. Robb, R. Carnap, H. Reichenbach, and H. Mehlberg.[12] Guided by their intuition or philosophical premises, they selected their axioms to give space-time the "most sensible" flow of time and other properties. But not until Gödel's discovery of a solution with closed time-like curves did researchers realise that it was not worthwhile ruling out certain possibilities a priori; instead as many solutions to Einstein's equations should be found as possible, and studied from the point of view of their global properties. This new trend helped to devise the global methods for the examination of space(-time) and launched a new style in differential geometry, different from the traditional practice. Once a series of particular solutions have been analysed it is possible to formulate general rules, and then to set about finding proofs for them. This is how many groundbreaking theorems have been arising. One of the first results of this approach was R.W. Bass and L. Witten's proof of the theorem which says that every compact space-time contains a closed time-like curve.[13] This was followed by a tide of further results. Brandon Carter systematised them in an extensive paper.[14] The crowning achievement in this line of research was the proof of the celebrated theorems of the existence of singularities (see the previous chapter). Global methods have become well-established both in relativistic physics as well as in pure geometry. The fountainhead giving rise to this new style of thinking was Kurt Gödel's work on a universe with closed time-like curves.

3. GOTT AND LI'S SUGGESTION

Gödel's solution not only launched new research methods. It also provided an opportunity for reflections of an ideological character. As we saw in the introduction to this chapter, there was no dearth of ideas before, either, to raise up the ideology of closed time to the rank of "the ultimate solution," but now an opportunity opened up to turn this ideology into a "self-explanatory" cosmological model. An example of work heading in this direction is the extensive paper by J. Richard Gott and Li-Xin Li under the suggestive title "Can the Universe Create Itself?"[15] Let's take a closer look at their idea.

Gott and Li wanted to make use of the "remarkable" property of the general theory of relativity – its admission of solutions with closed time-like curves, but they were aware of the difficulties this property implied. Closed time meant problems with causality, often expressed in the question what would happen if someone who made use of the time-loop killed his father before his own birth. In physics this provocative question translates into computational problems connected, for example, with the expected behaviour of a solution to a differential equation given its initial conditions. Moreover, there are no experimental clues that in our universe time is a closed loop. On the contrary, the scientific reconstruction of its history, by now based on numerous observations, makes up a coherent cosmic history with a linear time-scale stretching back to the first moments following the Big Bang. Gott and Li are too experienced as cosmologists not to know of all these difficulties. That is why the model they proposed was far more sophisticated than the simple models with a closed history.

To obviate a beginning, Gott and Li assume that "the early universe contained a region with closed time-like curves." Such a universe is neither eternal, nor has a beginning. Every event that happens in it has an event which preceded it, but the question which event was the earliest is as meaningless as asking which is the easternmost point of the Earth's surface. But at a certain moment this spell of "dodged history" came to an end, and now history is proceeding in a one-way direction towards the future. However, it is not enough to juxtapose the "early" period with a closed history with the "later" period of linear history. To avoid a variety of pathologies with causation, the region with closed time-like curves must be separated off from the later one-way history of the universe. Relativistic cosmology offers such a possibility. Due to the maximum, finite velocity at which physical interactions can propagate in space-time, there may exist regions with which no communication whatsoever is possible. No physical interaction can "get through" from one such region to another, and we describe the situation by saying that they are separated from each other by a Cauchy horizon. Gott and Li applied this mechanism to save the later history of the universe from the causal anomalies generated by closed time-like curves in the early universe. The two periods are separated off from each other by a Cauchy horizon.

But that was not the end of the new model's problems. Papers were published which showed that for Cauchy horizons of this type in space-times with closed time-like curves certain mathematical expressions describing the distribution of matter tended to infinity.[16] So it was possible to by-pass the pathologies connected with closed time only at the expense of bringing in other pathologies (the introduction of infinity). Gott and Li have challenged this finding. According to them in such situations it is possible to remove the "tendency to infinity" by finding a solution exactly reproducing conditions[17] which have already occurred before. A proposition which is enjoying considerable popularity nowadays are the so-called inflationary models, with the early universe expanding at a dramatically rapid rate – increasing its volume up to 10^{30} times or more in a fraction of a second! Although there are no observations to confirm them, such models have been well received in contemporary cosmology, because they resolve several theoretical difficulties.[18] Gott and Li argue as follows: let's assume that "in the beginning" there was an inflationary model; then the small volume of space-time was inflated to a gigantic size. If in that huge, inflated universe a small sub-region happened to occur with the same conditions as those in the initial, small volume, then closed time could have occurred without the need for infinity to come into play. "If that happened the universe could be its own mother."

It has to be admitted that the construction Gott and Li have presented is intricate, but still incomplete. An accumulation of computations and particular examples is not yet a full cosmological model. To construct a full model it is necessary to examine its stability and determine the set of initial conditions (to form the space of all initial conditions) which yields such a solution. If the solution calls for highly specific initial conditions it is in need of explanation itself, rather than serving to explain. And above all there is the question of whether the given solution corresponds to reality, viz. can it be verified by observed facts. The model proposed by Gott and Li cannot claim an answer in the affirmative to this question.

4. CAUSALITY AND TIME

The solution to Einstein's equations found by Gödel, and subsequently many other solutions with similar properties, have proved irrefutably that worlds with closed time are within the realm of possibilities afforded by the general theory of relativity. But it is still an open question whether such solutions are physically realistic, or whether they are a purely theoretical possibility; or, to put it more precisely, whether there exist any laws of physics which would make the closing up of time-like curves impossible. Such laws would serve as selection rules admitting only of solutions with unclosed time. Do selection rules of this kind exist?

As we have already said, other "strange behaviours" apart from time-loops may occur in space-time, giving rise to a variety of causal pathologies. For instance, almost closed time-like curves may occur in space-time. These do not give rise to any problems with causality in themselves, but a slight disturbance in the gravitational field – and this can never be ruled out – may bring about their closure, which would in turn entail a "causal disaster." Brandon Carter[19] has compiled an immense hierarchy of such instances of pathological behaviour. The question arises whether a general rule can be formulated for the elimination of all of them. It turns out that it can. But before we formulate it, we shall have to recall a few basic concepts from the geometry of space-time.

As is well-known, in the theory of relativity time-like curves represent the histories of particles which have a rest mass (which we usually call material particles). But apart from time-like curves there are also null curves (also known as light curves), which represent particles with zero rest mass, i.e. photons. Time-like and null curves together are referred to as causal curves. The occurrence of closed null curves would also give rise to a variety of pathologies, e.g. by using a photon we would be able to send a message back into the past. So in order to exclude all the causal pathologies we should also take into account the behaviour of the null curves, in other words speak of the causal curves.

Let's assume that we have a space-time in which there are no closed causal curves, but we also want to protect it against any causal pathologies which might threaten it. The rule is fairly obvious. What we have to do is impose a requirement that there should be no disturbance of the gravitational field, however small,[20] capable of bringing about the occurrence of closed causal curves. Space-times in which this rule is observed are known as *stably causal* space-times. As we have said, this rule seems fairly obvious, but the proof that in stably causal space-times there are no unwanted pathologies is not at all straightforward.

We should expect that if there are no problems over causality in stably causal space-times, then there should be no trouble, either, with the time record for the history of the universe. To verify this assumption we first have to have a "theoretical clock" available to measure the universe's time.

Let's take a look at any clock that we use in everyday life. It will be a device which assigns a real number to each particular moment. For example, as I am writing these words my wristwatch is showing an assignation of the numbers 9 and 36/60 in the conventional units known as hours and minutes. If I now recall that I and my watch are marking out a time-like curve in space-time, then I may designate any function continuously [21] increasing along a time-like curve as a clock.[22] If such clocks exist along every time-like curve in the given space-time, then there exists a global time in that space-time. Or strictly speaking, we say that a global time (also known as cosmic time) exists in the given space-time if in that space-time there exists a continuous function with real number values continuously increasing along every causal curve.

Our attention to these somewhat pedantic definitions is rewarded with an elegant theorem:[23] in a space-time there exists a global time if and only if the given space-time is stably causal.

5. PHYSICS AND GLOBAL TIME

We now know how to rule out causal pathologies and along with them lay the ghost of closed time: the requirement is that the space-time must be stably causal. But are there any additional physical reasons to justify this postulate? Again the answer is yes, but in order to appreciate it we have to consider the problem of measurements in physics.

As is well-known, every measurement is subject to error; there are no perfect measurements. Let's assume that a theory of physics T predicts that a certain specific reading for a physical parameter q is to give a value q_o. To prove or disprove the theory, we carry out the measurement, but we may not expect to obtain a reading of exactly q_o. We shall consider the measuring experiment to have verified theory T if the result we have obtained for the reading lies in the range $[q_o - \Delta q_o, q_o + \Delta q_o]$, where Δq_o is appropriately small. If we obtain such a reading, we are entitled to say that the experiment has confirmed the theory to a good approximation.

But a certain condition must be satisfied for the entire procedure to make sense. Let's assume that a small disturbance of the conditions in which the measurement is taken gives rise to very diverse results. The box of errors $[q_o - \Delta q_o, q_o + \Delta q_o]$ would then contain very many values of q, and we would not be able to say whether the experimental measurements had confirmed theory T, which had predicted the value q_o, or some other theory for which the predicted value would lie inside the box of errors. We must therefore assume that a small perturbation of the conditions in which the measurement is made gives rise to small changes in the results of measurement. This assumption is known as the assumption of *structural stability* of measurement, and is a (generally tacitly) adopted assumption of the experimental method. Without this assumption the experimental method would be groundless.

Measurements of time and space, viz. readings for time intervals and lengths in space, are some of the most important measurements in physics. The measuring-rod and the clock belong to the physicist's fundamental set of instruments.

Hence measurements of time and space should also be characterised by structural stability. But in the theory of relativity space and time are only aspects, observed from a certain reference frame, of space-time, which is independent of the choice of a frame of reference. Therefore, in compliance with the postulate of structural stability, a small perturbation in the structure of space-time may yield only small changes in the results of space-time measurements. But the structure of space-time is determined by the gravitational field, which in the general theory of relativity is the curvature of space-time,[24] in other words small perturbations in the gravitational field may give rise only to small changes in the results of space-time measurements.

Let us now consider a space–time in which there are no closed time-like curves. Perturbations of the gravitational field that produce closed time-like curves may not be called small. The postulate excluding such an occurrence, we recall, is known as the postulate of causal stability and is a necessary and sufficient condition for the existence of a global time in the universe. Therefore there is a strict correlation between the existence of global time and the possibility of measurements of space and time to be conducted, in other words the very possibility of physics as an experimental science. If the principle of stable causality were not valid in the universe, there would be no global time, small perturbations in the gravitational field could give rise to large changes in the structure of space-time, the principle of structural stability would not hold, and the experimental method employed in physics would be in jeopardy.

6. THE SPACE-TIME FOAM

Does all this mean that in the real universe time-loops cannot occur, since if they did, it would be impossible to conduct experimental physics? In the macrocosm we have been engaged in the practise of physics for quite a long time, and our level of success shows that the experimental method applied in physics is working very well, which in turn is a strong argument for the stability of the properties of the universe, in other words for the existence of global time. Global time operates at the macrocosmic level; while at the Planck level, until we obtain a complete theory of time, we shall have to be ready to admit a variety of possibilities. For instance, according to a fairly popular hypothesis, the closer we come to the Planck level, the more contorted space-time becomes, until at the Planck level itself it turns into a jumble of all the possible geometrical options. Various configurations of curves, including closed time-like curves, may occur in such a "space-time foam." Thus time also participates in the jumble of geometrical forms. Only as it proceeds to the higher levels does space-time gradually smooth out and a time measuring out cosmic history emerges.

However we should remember that the concept of a "space-time foam" is highly hypothetical, and the possibility of time-loops occurring in such a chaos of diverse configurations does not smack of an ultimate explanation.

Chapter 5

CONTINUOUS CREATION VERSUS
A BEGINNING

1. FROM THE STATIC TO THE STEADY STATE

One of man's ancient dreams is to build a *perpetuum mobile* – a machine which would work without the need to take in energy from without. The discovery of the second law of thermodynamics brought an end to such dreams: in an isolated system, although the total amount of energy is conserved (by the first law of thermodynamics), nevertheless it is dissipated and the machine's potential to perform useful work decreases all the time. But does the universe as a whole not fulfil the dream of the *perpetuum mobile*? Does it not ultimately provide some kind of explanation for its own existence? That it has always existed and will continue to exist forever. The early attempts in cosmology to accomplish this ideal failed. Contrary to his intentions, Einstein did not succeed in constructing a static model of an eternal universe. The universe is dynamic; it does not seem likely for changeability to last interminably; thus there looms a "ghost of the beginning." A solution was to come in the form of a perpetually oscillating cosmos, but this, too, turned out to be problematic in view of thermodynamics. The idea of a closed time is an alternative to a cosmology with a beginning, but one that replaces the latter with problems with causality, and even logic. All the indications are that if we wish to have a model of an eternal cosmos, we shall have to fit it out with some additional regenerative mechanisms. The first brave attempt of this sort was the cosmology of the steady state put forward in 1948 by Hermann Bondi, Thomas Gold, and Fred Hoyle. It expressed their reaction to the undeniable theoretical and experimental problems (the age

M. Heller, *Ultimate Explanations of the Universe*, DOI 10.1007/978-3-642-02103-9_5,
© Springer-Verlag Berlin Heidelberg 2009

of the universe) challenging the young relativistic cosmology; but from its outset it was also inspired by philosophical considerations and worldview. Its three designers found it particularly hard to accept the existence of a singularity at the beginning of "relativistic evolution." The theory's predictability broke down at the singularity – it was impossible to determine what happened before the beginning – and the capacity to predict is the fundamental feature required of any theory in physics.[1] Moreover, the singularity was too uncomfortably reminiscent of the concept of a creation of the universe, which all three scientists rejected on grounds of worldview.

There is a persistent habit of mind which suggests that an eternal universe must be static and unchanging. But must it? Could not a dynamic state coexist with eternity? It could, but the continuous dissipation of energy would have to be counterbalanced with some sort of "regenerating" mechanism. After Hubble's observations it was impossible to return to the idea of a static world. If the universe was to be eternal, it had to be a stationary system, viz. notwithstanding its variability it must always look the same. The density of matter decreases with increasing distance separating the galaxies moving away from each other; there was thus a need for a mechanism which would continuously restore the losses.

2. A NEW COSMOLOGY IS BORN

The theory of the universe in a steady state appeared in two versions: Bondi and Gold's, and Hoyle's. Initially Gold's idea of a continuous creation of matter was quite vague, but developed a more definite shape in the course of the three scientists' discussions. Gradually, however, they went their separate ways. Hoyle followed a more mathematical reasoning and tried to reconcile the concept of the creation of matter with the formalism of the general theory of relativity; while Bondi and Gold took an approach in opposition to relativistic cosmology, building up their model from scratch. Emphatically, neither Hoyle nor Bondi and Gold rejected the general theory of relativity as a theory in physics; they were only against its application in cosmology which they said was an unwarranted extrapolation. In effect two independent papers were produced, and despite the rivalry between their authors, by a strange coincidence both were published in the same issue of *The Monthly Notices of the Royal Astronomical Society*.

Hoyle was faster than his competitors and his article was ready much earlier, but it was turned down by the editors of two scientific journals. The prestigious British *Proceedings of the Royal Physical Society* gave the postwar shortage of paper as the grounds for its rejection; while the editors of the American *Physical*

Review wanted Hoyle to shorten the article, which he refused to do. In outcome it appeared in the *Monthly Notices of the Royal Astronomical Society.*[2] Originally Hoyle did not want to publish in this journal, which was edited by British astronomers, since he was apprehensive of their reaction to his unconventional ideas. But in fact quite the opposite happened, as the secretary of the British Astronomical Society at the time, who decided as to what was to be published, was William McCrea, a supporter of the hypothesis of the continuous creation of matter. The same man had already earmarked Bondi and Gold's article for publication.[3] They had not been trying their luck with other journals.[4] Thereby a great controversy was launched between the cosmology of the steady state and relativistic cosmology, and for the next two decades was to dominate developments in the science of the universe.

3. BONDI AND GOLD'S UNIVERSE

A summary of Bondi and Gold's model is to be found in Hermann Bondi's once highly influential textbook of cosmology:[5]

> *The fundamental assumption of the theory is that the universe presents on the large scale an unchanging aspect. Since the universe must (on thermodynamic grounds) be expanding, new matter must be continually created in order to keep the density constant. As ageing nebulae drift apart, due to the general motion of expansion, new nebulae are formed in the intergalactic spaces by condensation of newly created matter. Nebulae of all ages hence exist with a certain frequency distribution.*

The principal assumption in the model of the steady state is that "the universe viewed globally does not change." The authors of this model call this the perfect cosmological principle. It differs from the (ordinary) cosmological principle applied in the Friedman-Lemaître cosmology in that it assumes that the picture of the universe is independent not only of the observer's position in space (as in the ordinary principle), but also of the point in time of his observation. A large part of Bondi and Gold's argument boils down to propaganda on behalf of the perfect cosmological principle.

Copernicus taught us that the Earth does not occupy a special place in space. Why should it have a special place in time? Cosmology is based on the assumption that the same laws of physics are valid throughout the entire universe. If, in accordance with relativistic cosmology, we assume that at the beginning of its evolution the universe experienced a superdense phase, then we can hardly expect the same laws of physics that we have today to apply in densities of the order of 10^{93} g/cm^3. But "if the universe presents the same aspect to every fundamental observer, wherever he is and at all times, then none of these difficulties and doubts arises."[6]

Of course a static-state universe, in which nothing changes, obeys the perfect cosmological principle. But it does not conform with what is observed, since in a world that is static there would have to be thermodynamic equilibrium, since there are no changes. That is not what we observe. There are large temperature differences in the universe, and we ourselves, living organisms, are systems in states which are far from equilibrium. In other words, according to Bondi, the perfect cosmological principle, together with observation and the laws of thermodynamics, shows that the universe is not in a static state. Therefore it must be either expanding or shrinking. But

In a contracting universe the Doppler shift leads to a disequilibrium in which radiation preponderates over matter, whereas the opposite is true in an expanding universe. Accordingly, the steady-state theory, alone amongst all theories, deduces the fact that the universe is expanding from the local observations of thermodynamic disequilibrium.[7]

For the steady-state model the observations of the red shift in galactic spectra merely confirm that the theory's deductive reasoning is right.

But the agreement of the perfect cosmological principle with the observed expansion of the universe can only be upheld at the cost of the assumption that matter is continually being created in space, so as to maintain a constant mean density throughout the universe, despite its expansion. Bondi stresses: "It should be clearly understood that the creation here discussed is the formation of matter not out of radiation but out of nothing."[8]

Of course, the creation of matter understood in this sense is in breach of the principle of the conservation of energy. Bondi and Gold are well aware of the fundamental role this principle plays in physics, but they emphasise that what is really important in physics is agreement with what is observed, "there is, however, no observational evidence whatever contradicting continual creation at the rate demanded by the perfect cosmological principle,"[9] which requires that a mass equivalent to an atom of hydrogen be created in every litre of volume at a mean rate of once in 5.10^{11} years – and there is no experiment sensitive enough to detect such an amount.

The perfect cosmological principle turned out to be a powerful enough assumption to allow for a determination of the geometry of the universe without recourse to gravitational field equations, which Bondi and Gold could not use. The steady state postulate almost immediately leads to a conclusion that the curvature of space should be zero,[10] and that the expansion must proceed at an exponential rate,[11] which effectively gives de Sitter's space-time. In the relativistic cosmology de Sitter's model is empty, but this is an outcome of the field equations, which imply that the density of matter must be zero. In Bondi and Gold's version of the steady-state cosmology there are no field equations, so there is no need for this conclusion to hold.

As we see, in Bondi and Gold's model everything is an elegant outcome of the initial assumptions. But even the most elegant outcome would not have impressed anyone if the model could not claim to make any empirical predictions. The fact that it could earned it a considerable degree of authority, despite initial reluctance. Its predicted observations were also an outcome of the perfect cosmological principle. New galaxies were appearing to replace those which were moving away, at a rate sufficient to keep mean galactic density constant. Moreover, the model predicted that there would be a uniform mean distribution of young and old galaxies in space; somewhat later the statistical distribution of young and old galaxies was determined for the steady-state model. This prediction differed essentially from the predictions of relativistic cosmology, according to which young galaxies were expected to be systematically further away than old galaxies, since in making observations at increasing distances we would be looking at a universe younger than what it was now, and there could be no old galaxies in a young universe. In the years immediately following the publication of the steady-state model the verification of the predictions made by the two theories was beyond the reach of astronomical observation, and the debate continued on the basis of theoretical arguments and each side pointing out their rival's weak points.

Without doubt, one of the weak points of relativistic cosmology was the problem of the age of the universe. On the basis of Hubble's law and the available data for the red shift the age of the universe was estimated at about 2 billion years, while estimates for some of the rocks on Earth, meteorites and stellar systems gave values of up to 5 billion years. The creators of steady-state cosmology did not fail to turn this argument to their advantage. A steady-state universe has an infinite age, of course, and there is no clash with any other estimates on a time scale.

4. HOYLE'S UNIVERSE

Hoyle was more conservative in his revolutionary ideas than Bondi and Gold. He wanted to preserve the conceptual framework of relativistic cosmology as far as possible, departing from it only at the point where the introduction of the hypothesis of the continual creation of matter called for it. In compliance with his philosophy he thought that in this way he would maintain all the advantages of a cosmology based on the general theory of relativity while avoiding the conceptual and observational problems associated with it.

As we know, there is a local principle of conservation of energy built into the general theory of relativity. To infringe it Hoyle introduced into the field equations a new tensor term, which he called the creation tensor. The creation tensor was inserted in the place of the term with the cosmological constant, which does not appear in Hoyle's equations. The creation tensor's mathematical properties are similar to those of the cosmological constant component, except that it does not obey the principle of conservation of energy. Thanks to this Hoyle's version of de Sitter's solution, which is "empty" in relativistic cosmology (viz. the density of matter has a zero value), is filled up with matter which is continually being created. Hoyle also showed that this solution was stable. Hoyle's equations also have other solutions apart from de Sitter's solution, but it was chiefly de Sitter's solution that was the focus of Hoyle's attention and of the discussion that ensued. It is precisely in this solution that Hoyle's vision of the world is identical with Bondi and Gold's vision; despite the differences, or even controversies, that later emerged between these scientists, the two theories were later treated as just two variants of the same cosmology. But discussions with Hoyle's version were easier: with its more elaborate mathematical apparatus it could be more readily amended and improved, but it was also more liable to criticism in the form of specific objections. And it was Hoyle's version that found itself in the centre of the debate that soon emerged.

5. IN THE HEAT OF DEBATE

The cosmology of the steady state found itself in the limelight of British public opinion chiefly as a result of a series of radio broadcasts which Fred Hoyle made in the spring of 1949, which made him a media personality. Later the broadcasts were published in a book,[12] and helped to popularise the steady-state model outside the British Isles. Initially the transfer of the discussion to the popular forum generated additional opposition to the new ideas from British astronomers and physicists. But there were also counter-arguments of a more scientific nature.

It is not my intention to record all the developments in the debate between the protagonists of the steady-state cosmology and the adherents of relativistic cosmology. There is an excellent monograph on the subject by Helge Kragh.[13] I shall merely delineate a few of the episodes in it, relevant to the main subject of the present reflections – the search for "ultimate solutions" in cosmology.

The temperature of the debate suggested that there was more at stake than just the weighing up of technical arguments. Quite naturally the idea of rejecting the law of the conservation of energy as put forward by the creators of the steady state concept attracted a volley of criticism. The conservation of energy was a fundamental principle of physics, it was pointed out, and in addition comments were lavished on the notion of the creation of matter. Herbert Dingle remarked that the steady-state model was no more "scientific" than the relativistic models, since it was of no account whether one were to assume a "big creation" at the beginning, or an infinite number of "minor creations" going on continuously. For science both were a *deus ex machina*.[14] Hoyle, of course, did not agree with this. He wrote

> *. . . I cannot see any good reason for preferring the big bang idea. Indeed it seems to me in the philosophical sense to be a distinctly unsatisfactory notion, since it puts the basic assumption out of sight where it can never be challenged by direct appeal to observation.*[15]

Nonetheless he had to come to terms with the fact that in his model, too, matter simply appeared out of the blue, and all the "mechanisms of creation" he referred to related to the physical properties of matter that was already in existence. His claim that "the creation field" was generated by the matter present in the universe remained in the realm of purely philosophical speculation.

By way of commentary I shall refer to the passage quoted from Hoyle. Kragh remarks that this was the first occurrence of the phrase "the big bang" in print. Most probably Hoyle must have used it earlier in his oral statements. He applied it ironically, to discredit the rival theory.

Hoyle also introduced antireligious motifs into the debate. He also attacked "Marxists and materialists," but made use of their arguments, seasoned with a dash of Positivism, against the Christian concepts of the creation and immortality of the soul.[16] This made the atmosphere of the controversy even hotter and certainly helped to spread news of it. In this connection George P. Thompson, holder of the Nobel prize for physics who had experimentally confirmed the existence of matter waves, wrote:

Probably every physicist would believe in a creation [of the universe] if the Bible had not unfortunately said something about it many years ago and made it seem old-fashioned.[17]

With time the worldview controversy associated with steady-state cosmology calmed down and the debate became more scientifically oriented, especially as advances in the technology of astronomy and radioastronomy made the prospect of confronting at least some of the predictions of the steady-state cosmology with observational data more and more of a reality.

6. THE DEMISE OF THE COSMOLOGY OF THE STEADY STATE

While the debate was still going on whether the universe was in a steady state or whether it was subject to evolution on a grand scale, distinct progress was being made in work connected with the general theory of relativity. More and more evidence was accumulating to confirm it as a first-class theory of physics. In 1960 Pound and Rebka were the first to successfully carry out a laboratory test of the general theory of relativity. They applied the Mössbauer effect to measure the change in the photon frequency of gamma radiation due to the difference in the Earth's gravitational field over a height of 22.6 m (the height of the tower on the Harvard University campus, where the experiment was conducted). Physicists were certainly impressed by their result. At the same time advances were being achieved at a rapid rate in the theoretical work on the general theory of relativity, which started to exert an impact on the development of mathematics. The geometrical methods devised for relativistic physics were gradually entering the realm of abstract mathematics concerned with modern differential geometry. All of this was bringing a change of atmosphere and making the cosmology of the steady state, which was in opposition to Einstein's theory of gravitation, lose its ground.

An even bigger contribution to this process came from the advances made in observational techniques in astronomy and radioastronomy and the parallel progress in relativistic cosmology, which was partly stimulated by these advances. From 1948 on George Gamow and his team were working on a new scenario for the processes which had occurred in the young hot universe. Gamow's scenario was based on a knowledge of contemporary nuclear physics, and launched a series of projects to determine the "chemical composition" of the universe. From the very outset Gamow tried to find confirmation for his ideas in observational research to determine the frequency of occurrence in the universe for the nuclei of particular chemical elements. Soon Hoyle, along with Eleanor and Geoffrey Burbidge and William Fowler, announced a rival theory of nucleo-genesis, according to which the nuclei of the chemical elements were created not in a hot Big Bang, but in the interior of massive stars. The motivation behind this postulate was undoubtedly an attempt to neutralise the advantage enjoyed by relativistic cosmology thanks to the results obtained by Gamow's group. Both parties initiated intensive research programmes in the field which later came to be called cosmic nucleosynthesis. The results these rival projects achieved were a surprise for both teams. It turned out that the conditions prevalent shortly after the Big Bang were indispensable to produce all the hydrogen nuclei, about 70% of the helium, and small amounts of a few of the other light chemical elements extant now in the universe, which was in agreement with Gamow's theory. But the nuclei of the rest of the chemical elements were shown to be produced even now in the interiors of massive stars – as predicted by the theory proposed by Hoyle, the Burbidges and Fowler.[18]

Another argument against steady-state cosmology was provided by the development of radioastronomy. Thanks to progress in observational techniques, still in their infant stage at the time, it became possible to compile more and more accurate catalogues of radio-sources, which in turn facilitated the carrying out of a variety of tests for cosmological theories. The most promising test was the count of radio-sources per unit solid angle, not exceeding a certain luminosity (viz. the flux density at a given frequency). If the universe is in the steady state, the graph for number of radio-sources versus luminosity should be a straight line with a fixed gradient (a gradient of -1.5 on a logarithmic scale). The first test of this kind was conducted in 1955 by M. Ryle and P.A.G. Scheuer, and their result suggested a disagreement with the predictions of steady-state cosmology. Soon more work followed, with results more and more in line with each other and showing that the radio-source count increased with increasing distance. This would indicate that the younger the universe had been, the higher the density of radio-sources in it, therefore it could not be in the steady state. In 1963 W. Davidson and M. Davies wrote an article summarising these results. Their main conclusion was that the results hitherto obtained in radioastronomy could not be explained by steady-state cosmology.[19]

The discovery of quasars – strong sources of radiowaves identified as optical objects similar to stars – was yet another challenge to steady-state cosmology. After the first inconclusive results, measurements for their red shifts began to bring evidence against a steady-state universe. To avoid such a conclusion the theory's defenders put forward hypotheses of the "local derivation" of quasars, according to which quasars were not "at cosmological distances," but were associated with certain exotic phenomena in our relatively close astronomical neighbourhood. However, the influx of new data made such hypotheses less and less plausible.[20]

Today the general consensus is that the final blow to steady-state cosmology was administered in 1965 by the discovery of microwave background radiation, although Helge Kragh is of the opinion that it was a blow delivered to a theory already in its death throes.[21] Microwave background radiation was discovered by Arno Penzias and Robert W. Wilson, and interpreted by Robert Dicke and his collaborators as the remnants of the Big Bang which initiated the current phase of cosmic evolution. The existence of this radiation had been predicted in the late 1940 s by George Gamow, who together with his team had determined its expected properties. It was to be an isotropic (viz. independent of direction) black body radiation at a temperature of a few degrees Kelvin. However, Gamow's prediction had been forgotten, and Dicke and his team at Princeton rediscovered it in their theoretical work.[22] Penzias and Wilson's observations confirmed all the theoretical expectations to a good degree of accuracy, and later measurements made the accuracy even sharper.

After the discovery of background radiation the popularity of steady-state cosmology fell dramatically. Even Hoyle was inclined to admit that the observed data indicating that the world was subject to global evolution were too serious to ignore and insist on the concept of the steady state. But he was still reluctant to acknowledge relativistic cosmology with the singularity at the beginning of evolution. He persisted with his claim that any "beginning" whatsoever contradicted the principles of scientific methodology, and kept coming up with a series of new variations of a theory with no Big Bang.

The demise of the steady state cosmology is an interesting case for the philosopher of science. As Kragh writes, it did not disappear overnight and for good; there was no "ultimately decisive" discovery, either in the theory or in the observations, to falsify the theory.

Rather, the controversy faded out in the sense that the now standard hot big-bang model became the nearly undisputed new paradigm in cosmology, and the new generation of cosmologists stopped worrying (or even knowing) about the steady-state theory.[23]

Philosophers of science hold that there is no such thing as an *experimentum crucis* (critical experiment) capable of disproving any given theory once and for all. A theory that is losing ground may always be modified and kept up by supplementary hypotheses. That is precisely what the supporters of steady-state cosmology did, trying to salvage it, but gradually their ranks crumbled away. The staunchest were Hoyle and his collaborator Jayant V. Narlikar,[24] but soon they found themselves out on a limb. It is an indisputable historical fact that the discovery of background radiation was the experiment which sealed the fate of steady-state cosmology. Even if it was not an *experimentum crucis* in the sense understood by the philosophers of science, in conjunction with the other observations indicative of an evolving universe it proved an obstacle which the theory of the steady state did not manage to overcome.[25]

7. CREATION AND VISCOSITY

Looking back in retrospect at the history of steady-state cosmology it is hard to avoid the impression that it was an *ad hoc* hypothesis, called into being precisely for the purpose of removing from cosmology the "ghost of a beginning" in the sense not of a technical problem in cosmology but of an attempt to arrive at an ultimate explanation of the universe. The steady-state theory survived for almost two decades only because it had some observational tests at its disposal which were relatively easy to conduct. These tests were carried out, and the theory disclosed its weak points. Compared with *ad hoc* hypotheses, genuine scientific theories are characterised by being aggressive in a certain sense: they tend to annex ideas which are not so forceful but in a way attractive though lacking in solid foundations, and incorporate them into their own models and techniques. This proved true in the case of relativistic cosmology and steady-state cosmology.

The general theory of relativity is one of the most important theories in contemporary physics, linked by a variety of connections with other theories of physics and together with them constituting a well-knit, albeit far from complete structure. Relativistic cosmology is the natural application of the general theory of relativity to the universe on its largest scale. The first cosmological models were based on a number of simplifications. One of these simplifications was the ignoring of the dissipation of energy. The collection of galaxies was treated as a dust the particles of which do not interact with each other, or as a perfect fluid in which there are no problems of dissipation. From Chap. 3 we remember that the first scientist to introduce dissipation processes into cosmological models was Richard Tolman. In models of the universe which obey the cosmological principle dissipative processes are introduced by adding to the equations terms responsible for bulk viscosity, also known as second viscosity. "First" viscosity is associated with interlayer friction, but this kind of viscosity cannot occur in isotropically expanding models (viz. observing the cosmological principle), since there is no interlayer friction in expansion of this kind. However, bulk viscosity, which is associated with the rapid expansion of a fluid, may of course occur in an expanding universe. From Chap. 3 we may recall the surprise when it turned out that on taking bulk viscosity into account in the oscillating model there was an increase in successive cycles of the oscillation. But we would expect the oscillations to diminish due to the dissipation of energy, as they do in classical physics. However, usually when we consider dissipation processes in classical physics we are talking about isolated systems, viz. ones which do not exchange energy with their surroundings. Strictly speaking, in the general theory of relativity we cannot, even in principle, construct an isolated system. We might perhaps cut off the supply of energy from beyond the system, but we cannot "switch off" the gravitational field which, being related to the space-time curvature, is stored in the geometry of space-time itself, and therefore penetrates all insulators. As the calculations show, processes involving bulk viscosity may draw energy from the curvature of space-time. This was precisely why Tolman's calculations showed that the cycles of an oscillating universe were not subject to damping down, but on the contrary – to an increasing amplitude. The mechanism for the production of energy from the curvature of space-time is responsible for the existence of many other solutions to Einstein's equations involving bulk viscosity apart from Tolman's "increasing cycles."[26] They include solutions in which the production of energy is exactly counterbalanced by the loss in density caused by the recession of galaxies (in other words we have steady-state solutions). One of them turns out to be exactly the same as the solution Hoyle found for his "creation field" equations. From the mathematical point of view this relativistic model involving second viscosity and Hoyle's model involving the creation of matter are indistinguishable from each other.[27] Thus the same solution admits of two different interpretations: one the standard lodged within an efficient theory of physics, and the other an *ad hoc* device, put forward effectively with only one aim in mind: to eliminate the problem of the beginning. Observations came out in favour of the stronger of the two.[28]

From the historical point of view the discovery of viscous models made no contribution to the downfall of steady-state cosmology. By the time the first papers on viscous cosmology were being published the steady-state theory had been out on the distant margins of cosmological research for quite a while already. Nevertheless, a moral may be drawn from the whole story for the focus of our attention – the scrutiny of ultimate explanations in cosmology. The moral is that ultimate explanations should be constructed on the basis of well-founded physical theories rather than on *ad hoc* ideas. This does not mean, of course, that in future the quest for ultimate explanations will not bring any profound revolutions in ideas. Quite on the contrary, we should be expecting such revolutions, but the chances of really profound revolutions are much greater when they are derived from problems in the mainstream of science.

Chapter 6

~

SOMETHING ALMOST OUT OF NOTHING

1. THE HORIZON PROBLEM AND THE FLATNESS PROBLEM

*A*n interesting, frequently overlooked point is the fact that some of the ideas associated with the steady-state cosmology turned out to be longer lived than the theory that produced them. The steady-state theory disappeared from the scientific scene as a serious rival of relativistic cosmology, but years later certain ideas associated with it revived in another form and as it were on a different level, sometimes with the application of different, more sophisticated concepts. The ideas concerned are both the "generation of matter", as well as the steady state, though not within our own universe but in an infinite set of universes. As we recall, the notion of a multiplicity of universes was put forward by Hoyle in his attempt to salvage the steady-state theory. The new cosmology in which these ideas re-emerged was inflationary cosmology, which appeared in an effort to deal with certain theoretical problems encountered by relativistic cosmology, which was otherwise making dynamic progress.

From the very outset measurements of the microwave background radiation had been indicating that at the time when this radiation ceased to interact with other forms of matter – and according to the most recent data collected by the WMAP satellite that meant 380 thousand years after the Big Bang – the universe was extremely homogeneous: any disturbance in its density would have given rise to a deviation from the mean temperature of the background radiation. Subsequent measurements narrowed down the constraints determining the homogeneity of the young universe. Today we know that the temperature of the microwave background radiation is constant over the entire expanse of the sky to an

M. Heller, *Ultimate Explanations of the Universe*, DOI 10.1007/978-3-642-02103-9_6,
© Springer-Verlag Berlin Heidelberg 2009

accuracy of 10^{-5}, in other words that was the level of accuracy to which the young universe was homogeneous. The question arises why this was so. "Not very special" initial conditions will do to produce a chaotic universe; but to produce a universe with that degree of homogeneity a set of highly specific initial conditions is required. Our "sense of what is realistic" suggests the following solution: perhaps the initial conditions might have been "not very special," but presumably there must have been some kind of physical interaction to smooth out the originally chaotic universe. This line of reasoning seems appealing, but again there is a snag in the measurements for background radiation. As we remember, its temperature is virtually constant across the whole of the sky. Consider two points at opposite ends of this range. It may be easily calculated that the history of the universe was too short for the fastest physical signal, light, to join these points together. We say that the two points are separated from one another by a horizon. Therefore there is no physical interaction capable of evening out temperatures in regions separated from each other by a horizon. The standard cosmological models were not able to cope with the horizon problem.[1]

A second analogous problem is the flatness problem. According to the equations of standard relativistic cosmology, the curvature of space is constant, but it may be zero, or take any positive or negative value between plus and minus infinity. The determination of the curvature by estimating the mean density of matter had for a long time been indicating that the space of our universe is almost completely flat, viz. that it has a curvature very near to zero. This has been confirmed to a high degree of accuracy by the latest measurements carried out by the WMAP satellite. Now the line of reasoning is similar to the one for the homogeneity of the universe. The only distinguished value for curvature between plus and minus infinity is zero. It is distinguished because it separates the negative curvatures from the positive ones. Why did the initial conditions "select" a value for curvature so close to the distinguished value? The standard models cannot answer this question.[2]

Both the horizon problem as well as the flatness problem[3] disappear if we assume that at an appropriately early stage of its evolution the universe underwent a rapid process of expansion, referred to as inflation. Up to that point the entire universe as observable today might have occupied a very small volume with no horizons splitting it, such that any physical processes occurring within that volume would have smoothed out all the "bumps." Only later was space inflated to the size of the universe as observed today. Before its inflation the universe might have had any arbitrary curvature, but following its inflation what we observe is only a "small," approximately flat sub-region of space which has an arbitrary curvature (on the principle that as long as it is smooth, any arbitrary surface, e.g. a sphere, is locally flat).

The inflationary scenario was first proposed by Alan Guth in 1981,[4] to resolve the two above-described difficulties in standard cosmology, but soon it rose to the status of a new research programme. In this chapter we shall be concerned with those aspects of inflationary cosmology which are connected with the main subject of this book – the search for ultimate explanations.

2. THE MECHANISM OF INFLATION

The inflationary model makes the assumption that at a very early stage in its evolution the universe experienced a sudden (exponential) acceleration in its expansion, which was "superimposed" on the normal expansion of the standard model. The accelerated expansion was propelled by a scalar field φ with an equation of state rather exotic from the point of view of later evolution. In this equation it is assumed that the pressure p of the "cosmic matter" is equal in magnitude to its density ρ but takes the opposite sign, viz. $p = -\rho$ (in units for which the speed of light $c = 1$). The function for the potential energy $V(\varphi)$ of the scalar field φ also plays a significant role. The possible inflationary scenarios depend on the shape of this function. The equation of state and the shape of the function $V(\varphi)$ are selected such that the scalar field φ acquires the properties of a "repellent gravity". The region of space over which the scalar field has these properties, called the "false vacuum" region, undergoes a very rapid expansion. In a fraction of a second the dimensions of this region may increase 10^{30}-fold (or more, depending on the exact scenario). We say that this region is in a "false vacuum" state. After a time this state is transformed into particles and radiation, and the accelerated expansion reverts into the normal expansion of the standard cosmological model.

Guth's original inflationary model met with certain difficulties connected with the departure from the inflationary state. To avoid these problems, the nature of which we shall not go into here, the model was modified several times: by Linde,[5] then by Albrecht and Steinhardt,[6] and finally again by Linde,[7] who put forward the chaotic inflationary model.

Linde's last model made the most significant impact on the philosophical reflections which soon emerged in connection with the inflationary model. For technical reasons Linde proposed an idiosyncratic shape for the function of the potential $V(\varphi)$. It was appealing theoretically, but required special conditions indispensable to start the inflation. How could such special conditions be justified?

Linde put forward a hypothesis that before the onset of inflation the world was in a chaotic state, viz. the physical fields assumed various values at various points in a random, "chaotic" distribution. Most of these fields were in the most probable states, which did not lead to inflation, but here and there conditions with a low probability of occurrence were extant, initiating the inflationary phase in the given region. These regions were inflated to huge sizes, whereas regions with no inflation remained microscopically small in comparison. Soon the universe was dominated by the inflated regions.[8] Each of these regions may be regarded as a separate universe, evolving independently of other similar universes. Moreover, in each of these universes the production of new inflated domains and the generation of new universes may recur. Linde's scenario made a significant contribution to the specific philosophical vogue that was soon to come for the multiverse concept, viz. speculations as to the existence of "all possible universes."[9]

The initial versions of the inflationary scenario combined the phase of the universe's rapidly accelerated expansion with the Grand Unifying Theories (GUT). Before the period of inflation three of the fundamental physical forces, strong nuclear, weak nuclear, and electromagnetic, were believed to have made up one force (the gravitational force had split away earlier); usually this process is considered in reversed time, and that is why we speak of a unification of the interactions. The separating off of the strong nuclear force from the other two was referred to as the phase transition associated with GUT. It was said to have occurred 10^{-35} s after the singularity, when energies of the order of 10^{14} GeV prevailed in the universe. This phase transition was believed to have initiated the inflation. The scalar field φ required by the concept of inflation was regarded as identical with Higgs' field, an essential constituent in the mechanism of the Grand Unification. However, it turned out that inflation combined with GUT would have produced too large a perturbation in the microwave background radiation compared with the perturbation actually observed. That is why now the inflation is no longer associated with the GUT phase transition, but considered separately; while the scalar field φ is no longer thought of as identical with Higgs' field, but is simply called the inflaton field or abbreviated to the inflaton.

One of the essential properties of inflation is the fact that while it was in progress energy density remained constant and assumed a value characteristic for the false vacuum. At the beginning of the inflationary phase there was a gigantic energy density, but as volume increased, the only way in which the density could be kept constant was by the creation of new energy.[10] Since, according to the inflationary scenario, it was from this energy that the universe's current "material contents" emerged, it is sometimes claimed that in principle all that is now observable arose out of nothing in the inflationary era.[11]

3. THE INFLATIONARY SCENARIO

Let's try to apply the mechanism described above to the scenario of the processes that were going on. We shall limit ourselves to a consideration only of the simplest version, which was later modified and amended many times and in various ways.[12]

Inflation appeared on the evolutionary scene soon after the era of quantum cosmology ended on "Planck's threshold," gravitation separated off from the other unified forces and a space-time governed by the laws of the general theory of relativity emerged from the Big Bang. The universe expanded in accordance with one of the Friedman-Lemaître models. Its material contents comprised hot plasma ("ordinary matter") and the inflaton field. Immediately after Planck's threshold was crossed the density of the plasma was of the order of 10^{93} g/cm^3 (the Planck density) and dominated the inflaton field to such an extent that the influence of the latter on evolution may be ignored. However, as the universe expanded the density of the plasma decreased (as R^{-4}), while the density of the inflaton remained unchanged. At a certain point the density of the plasma was equal to the inflaton density (in Guth's original model this occurred when the universe was 10^{-35} s old); at this time the strong nuclear interaction split away from the electroweak interaction. Subsequently the density of the inflaton started to predominate and the universe entered the inflation era. Its linear dimensions increased at an exponential rate. The plasma was rapidly diluted and its density reduced at an exponential rate. After a short time there was hardly any plasma left in the universe. But, as we know, energy density remained constant, therefore, in view of the rapid inflation by volume, energy had to be created.

There are several scenarios for the end of the inflationary era (this is still the theory's most delicate point). In all cases the energy of the inflaton transformed into the energy of the elementary particles present now in the universe. It is estimated that in the whole of our galaxy there may be just one proton or electron at the most derived from the pre-inflation era. On leaving the inflation era the universe was thermalised (heated up), and the rest of its evolution followed the standard model. The initial conditions for this evolution were determined by the physical processes which brought the universe out of the inflationary phase. What had come before inflation did not influence what came after.

The various models for the inflation envisage different times of its duration. If we assume that inflation lasted 10^{-30} s, then by the end of inflation the linear dimensions of the universe would have increased by a factor of the order of 10^{28}. It may be readily calculated that if before the inflation the typical size of the universe was of the order of 10^{-30} cm (viz. about a thousand Planck units), then after the inflation its typical size was of the order of 10^{-2} cm. A macroscopic size, nonetheless we are shocked by its "smallness." The shock should help us realise that when we are talking about inflation we are really very close to a "beginning."

4. SOME CRITICAL REMARKS

The inflationary scenario has a very strong presence in the cosmological literature, but we have to bear in mind that it is highly hypothetical and prone to a number of objections. We shall enumerate a few of the most important ones, after Gordon McCabe:[13]

First, according to what is known today in cosmology, the observable evolution of the universe may be explained in two ways: either by simply assuming the appropriate initial conditions, or by invoking a variety of physical processes which acted causally to bring about the evolution of the universe as it is today. The inflationary scenarios imply the latter, which seems the more appealing of the two ways. However, until we get a fundamental theory of physics we cannot be sure that the initial conditions responsible for the current state of the universe, acting "at the beginning" of its evolution, were not the outcome of some still unknown necessities which had nothing at all to do with inflation.

Secondly, if there was an inflation, then not all the material "contents of the universe," but *almost* all of them grew out of nothing during the inflation, since at the start of the inflationary phase there was already a certain "extant" energy density, which merely "proliferated" to keep the density constant.

Thirdly, the inflationary scenario in itself does not provide a guarantee that almost all the "material contents" of the universe grew out of nothing during the inflation phase. An inflationary universe may be either spatially infinite (non-compact), or spatially finite (compact). Only in the latter case would the universe have a finite volume, and almost all of its (non-gravitational) energy could have been created in the inflationary era. But if the former instance had occurred, then inflation could not have affected all of the universe and the matter beyond the inflationary area must have come about in some other manner.

Fourthly, inflation does not explain the origin of either space-time or the initial amount of energy.

And fifthly, an inflaton field is necessary for inflation to come about. In the current inflation scenarios the inflaton field is "manually inserted" into the equations. The justification of its existence by reference to mechanisms known from other branches of physics which should have been active in the early phases of the universe's evolution is still an open question.

Despite these difficulties, the inflation idea has become well-established in cosmology. It does explain several problems in standard cosmology, but itself requires a better foundation, as regards both theory and (especially) observational data. The concept of inflation has certainly not provided an "ultimate explanation" of the universe, but, as the latest history of cosmology shows, it has staked out a new path in the search for such explanations. Quite paradoxically, this has happened not so much thanks to its basic idea, but rather to a "side product," that is thanks to the fact that some inflationary scenarios postulate the existence of "other universes," different from the one in which we live and which we can observe, or even completely disconnected from it. This idea would return later in a variety of forms, to become one of the approaches in the search for "ultimate explanations." But before that happened, cosmology would be enriched with a number of brave new ideas.

Chapter 7

～

THE QUANTUM CREATION
OF THE UNIVERSE

1. FROM INFLATION TO CREATION

The inflation case has made it clear that if we want to approach an "ultimate explanation" we shall have to think of something more radical than just bloating up the size of the universe. That little "something" which triggers inflation calls for an explanation as well. But the inflationary mechanisms have drawn our attention to the vacuum problem. The quantum vacuum with which contemporary physics is concerned is not the metaphysical nothingness from which we should like to produce everything that exists (and thereby achieve the "ultimate explanation"), but it is the physical state of least admissible energy, and we suspect that it must have played an important part in the emergence of the universe from something more primordial than the states of the universe that modern physics is capable of describing.

The hypothetical "false vacuum" indispensable for the initiation and maintenance of inflation, as we saw in the previous chapter, is not the same as a "real" physical vacuum, which is defined as the global minimum of the potential energy function. In classical physics it is assumed that this minimum potential energy of the physical fields under consideration is equal to zero (as we know, we may select any point on the energy scale as our zero point). In quantum physics we cannot do this, since in accordance with Heisenberg's principle, an exact determination of the energy level (even if it were the zero level) would mean an infinite uncertainty of the time for the entire process. On combining this fact with other laws from the relativistic quantum theory we obtain a picture of the quantum

M. Heller, *Ultimate Explanations of the Universe*, DOI 10.1007/978-3-642-02103-9_7,
© Springer-Verlag Berlin Heidelberg 2009

vacuum as a container in which there is an "eternal storm" of various processes. In the quantum vacuum pairs of particles and anti-particles are continually being generated, only to be annihilated shortly afterwards. The quantum vacuum is not a static nothingness, but an ocean of fluctuating energy.

Could we not use the quantum vacuum to produce the universe? Of course it would not be the creation out of nothing of the universe that theologians speak of, but undoubtedly it would mark a step forward on the road to finding explanations that go further and further.

2. A UNIVERSE OUT OF THE FLUCTUATIONS OF A VACUUM

In the early 1970s Edward Tryon sent an article to the prestigious periodical *Physical Review Letters* on the emergence of the universe out of the quantum vacuum, but the editors rejected it as too speculative.[1] Tryon revised it and sent it to the no less prestigious *Nature*. We may safely say that his article was highly successful, launching a new path of research in the quest for the origins of the universe.

The idea itself was fairly simple. Something could arise out of nothing provided the process obeyed the law of the conservation of energy. This would be possible if that "something" had a total energy equal to zero: if, for example, the various energies in the "something" had different signs and cancelled each other out. Then the total energy before and after the "creation" would be equal to zero and the law of conservation of energy would be observed. We had already known for a long time that the total energy in the closed Friedman-Lemaître cosmological model was equal to zero, since the energy of the gravitational field was negative and exactly cancelled out the positive energy contained in the masses.[2] Could the closed Friedman-Lemaître universe arise out of a "zero" energy, in other words out of a void? In the deterministic classical physics this would be an impossibility, but it becomes possible if the initial state is a quantum vacuum. Let's imagine that a small particle is generated from the fluctuation of a quantum vacuum.

This generates a gravitational field which, by standard quantum-mechanical processes gives rise to the production of particles, which produce more grav-itational fields. . . and so on. Thus there is a sort of zero-energy conserving, fire-ball explosion away from the initial nucleation, and this is what can be thought of as a model for the Big Bang.[3]

Tryon's idea became rather popular. His model did not provide an "ultimate explanation," as it did not explain the origin of the quantum vacuum the fluctuation of which gave rise to the universe (Tryon spoke of a "pre-existing quantum vacuum"), but the concept of a vacuum seemed near enough to nothingness for the idea to inspire many researchers who took up this line of reasoning. Some of them found the combination of Tryon's idea with the concept of inflation particularly exciting. Three Belgian researchers, R. Brout, F. Englert, and E. Gunzig, proposed a model in which the matter emerging from quantum fluctuations had a large negative pressure, giving rise to an inflation scenario.[4] A number of other researchers pursued the same line of enquiry. Tryon resorted to this idea to underpin his model. Whatever happens in a quantum vacuum is determined by the law of probability; therefore the chances of a small universe emerging from it are much higher than the chances of a big universe emerging, but our universe is very big. But it cannot be ruled out that in the beginning it was very small and that inflation expanded it to a huge size.[5]

The idea that the negative energy of a gravitational field could cancel out the positive energy contained in the masses is certainly appealing and carries numerous consequences, but we must not turn a blind eye to the problems which it has to reckon with. It is textbook knowledge that in the general theory of relativity there are serious problems with the definition, independently of choice of coordinates, of the localised energy of a gravitational field. Hitherto the only case where such a definition has been successfully obtained was for an asymptotically flat space-time, viz. one which admitted the assumption that at "infinity" (that is at a sufficiently remote distance away from the observer) the gravitational field was weak enough to be ignored. Such a situation definitely does not correspond to any of the more realistic cosmological models. In the general case it has been an open issue, but many researchers have been inclined to conclude that the concept of the total energy of the universe is meaningless. If so, then the whole of Tryon's construction is unfounded.

There is one more snag to the idea: the universe is not just its "material content," but also its space-time. We may assume that according to Tryon's model the universe arises out of a pre-existing quantum vacuum and a pre-existing space-time, but the status of the space-time in this model is not clear. The concept of space-time is one of the tools of relativistic physics rather than quantum physics, but Tryon's concept does not even have the rudiments of a quantum theory of gravitation. Hence it can be no more than a prelude to, or inspiration for more advanced ideas.

The next stage was the attempt to "produce" the universe, along with its space-time, out of "nothing," on the assumption of the existence of only the laws of physics. The general consensus among theoreticians was that these laws should combine quantum physics with gravitational physics. But since we still do not have a generally approved quantum theory of gravitation, a set of hypothetical assumptions had to be made regarding this issue. The model which became the best-known concept of "the quantum creation of the universe" was Jim Hartle and Steve Hawking's proposal (1983)[6]: a hybrid of two extremely hypothetical models for the quantisation of gravitation – a model based on the concept of the quantum function for the universe, and the integration over paths model. Before we present the Hartle-Hawking model, we have to turn our attention to these two constituent models.

3. THE WAVE FUNCTION OF THE UNIVERSE

There are several different approaches to ordinary quantum mechanics. For most issues in this branch of physics they are equivalent, only in applications concerning quantum field theory do some of them turn out to be more useful than others. But the fundamental differences between the various approaches only become conspicuous when we attempt to apply them to the quantisation of the gravitational field. This is where the different strategies arise in the search for a quantum theory of gravitation. To achieve this diverse authors have been trying to apply a variety of approaches to ordinary quantum mechanics.

The most familiar, textbook-style formulation of quantum mechanics boils down to a statement that in a certain space known as the configuration space a wave function (usually labelled Ψ) is defined containing all the available information on the quantum object under consideration (e.g. an electron). The wave function must obey the differential equation governing its evolution. In standard quantum mechanics the wave equation is the well-known Schrödinger equation. Usually its solution and the interpretation of the results obtained mark the end of the theoretical part of the problem.

Difficulties crop up as soon as we try to transfer this method to the quantisation of gravitation. Above all the configuration space turns out to be very complicated. The contemporary theory of gravitation, that is the general theory of relativity, is set in a four-dimensional space-time, but space-time is not a quantum object capable of taking part in a game of quantum probabilities. To transform it into such an object it has to be resolved into all possible three-dimensional spaces. This is a complicated procedure, since in attempting such a resolution it is very easy to produce many copies of the same three-dimensional space differing only by having a different mathematical description. A lot of effort was expended before theoretical physicists learned to carry out this procedure in the right way. But this was not the end of the problems relating to the construction of a configuration space. The three-dimensional spaces have to be equipped with all the possible configurations of geometry and physical fields.[7] Only when we have a configuration space constructed in this way may we determine the wave function of the universe on it. But this is where the really big conceptual problems start. What does the "wave function of the *universe*" mean?

In the 1920s, when Schrödinger brought the notion of the wave function of the electron into quantum mechanics, he misinterpreted it himself, and quite a long time had to pass before physicists agreed on its probabilistic interpretation. According to this interpretation those properties of the electron for which the wave function has the biggest value have the greatest probability of being achieved. This interpretation has to be transferred in some way to the wave function of the universe. Every three-dimensional space with determined fields represents a possible state of the universe. There is an infinite number of these states. The wave function of the universe is determined over the space of all of these states. Those states for which the wave function has a bigger value have a higher probability of being achieved. The wave function should have the highest values for those states which describe a universe similar to ours – because that is what the real universe is like.[8]

The wave function for the universe should obey a differential equation similar to Schrödinger's equation. It is known as the Wheeler-DeWitt equation. Although it has an analogous role to Schrödinger's equation in quantum mechanics, it differs significantly from the latter. Schrödinger's equation describes the evolution of the wave function with time, but how can a wave function determined over all the possible states of the universe evolve? All the possible states of the universe do not exist in time. There is nothing with respect to which the wave function of the universe may evolve. Again a lot of time passed before physicists arrived at the right way to tackle that problem. The heart of the matter lies in the Wheeler-DeWitt equation. The wave function of the universe depends on various parameters characterising the possible states of the universe, and the Wheeler-DeWitt equation describes the changes in the wave function for the universe with respect to all those parameters. Time turns out to be a correlation between some of them. So there is no external time (external with respect to the universe) which can be used to determine the changes and rates of change in the universe. Time is an outcome of the internal relationships between the parameters characterising all the possible states of the universe. The Wheeler-De Witt equation plays the role of a co-ordinator, selecting a set of states out of all the possible states which lead to the emergence of "internal time."

This theoretical scheme is often referred to as the canonical quantisation of the general theory of relativity. Its chief assets are some interesting conceptual analyses which elucidate the nature of the difficulties encountered in diverse attempts to quantise gravitation. It has been developed as an independent research programme, but it constitutes only one of the two models on which Hartle and Hawking based their idea of the quantum creation of the universe. The other is the integration over paths model, which is frequently applied in quantum field theories.

4. PATH INTEGRALS

In this approach we are interested not so much in states as in transitions from one state to another. Let's consider two states S_1 and S_2 of a quantum system; we want to calculate the probability of a transition from state S_1 to state S_2. To do this we calculate all the possible paths in the configuration space from S_1 to S_2. Along each of these paths we calculate a certain integral (referred to as the action integral, in other words to each of the paths we assign a certain number, which is the result of the integration. Effectively we obtain a function defined on all the possible paths from S_1 to S_2. This function is associated with the probability of a transition of the quantum system from S_1 to S_2.

This method works very well in quantum field theories, but when we try to apply it to the theory of general relativity we are faced with serious problems. This is what Hartle and Hawking attempted. We shall take a closer look at their procedure.

We shall consider a transition from S_1 to S_2, just as we would for ordinary quantum mechanics, but now we shall be dealing with states of the universe. Each of these states is a three-dimensional space S with an appropriate metric tensor γ (which defines the geometry on S), and the appropriate physical fields φ. We shall adopt Hartle and Hawking's assumption that S is a closed space (like a three-dimensional sphere). We shall describe the initial state S_1 as the triple $(S_1, \gamma_1, \varphi_1)$, and the final state S_2 as the triple $(S_2, \gamma_2, \varphi_2)$.

The path from S_1 to S_2 is a sequence of "intermediary" states of the universe, in other words a sequence of closed three-dimensional spaces with appropriate fields γ and φ. Of course certain conditions must be fulfilled for smooth transition from one state to the next. The sequence of states traces a "tube" in the space of all possible states. States S_1 and S_2 are the boundary states of the tube. Now we have to consider all such tubes starting at S_1 and finishing at S_2 and calculate the magnitude (referred to as the propagator) which allows for the determination of the probability of a transition from state S_1 of the universe to state S_2 of the universe. The propagator is usually denoted by the symbol $K(S_1, \gamma_1, \varphi_1; S_2, \gamma_2, \varphi_2)$.

Unfortunately there are a number of conceptual and technical difficulties which complicate the solution of the problem. One of the most serious is the fact that in the general theory of relativity three-dimensional spaces "at a fixed point in time" have to fit into a four-dimensional space-time. As we know, in space-time the square of the time coordinate in the expression for the space-time metric takes the opposite sign with respect to the spatial coordinates. We say that the geometry of space-time is Lorentzian, not Riemannian (in the latter all the coordinates take the same sign). The difficulty is that in the Lorentzian case the calculations necessary for the computation of the probability of transition from state to state are generally impossible to carry out, for fundamental reasons.

To get round this difficulty Hartle and Hawking used a certain trick which is sometimes resorted to in ordinary quantum mechanics, in situations when the time coordinate t appears in equations. They multiplied t by the imaginary unit i, the square root of minus 1. This procedure made all the coordinates in the space-time metric assume the same sign and turned the Lorentzian space-time into a four-dimensional Riemannian space. In ordinary quantum mechanics a similar procedure is regarded as a trick in calculations, and after the calculations have been carried out the original sign is restored to the time coordinate. Hartle and Hawking assigned a fundamental meaning to this operation. They interpreted it as a mathematical expression of time at the basic level losing its properties of "the flow of transience" and turning into a fourth spatial coordinate.

Their next investment was the assumption that the propagator is the wave function of the universe, in other words

$$\psi = K(S_1, \gamma_1, \varphi_1; S_2, \gamma_2, \varphi_2).$$

This is where the programme of the canonical quantisation of gravitation meets the integration over paths programme. The wave function is a conceptual component of the former, and the propagator is a conceptual component of the latter. Moreover, Hartle and Hawking postulate that the wave function of the universe obey the Wheeler-DeWitt equation.

Now comes their most important conceptual innovation. Let's imagine that the initial state is the "empty" state, viz. $S_1 = \emptyset$. We shall now calculate the wave function

$$\psi_0 = K(\emptyset; S_2, \gamma_2, \varphi_2).$$

This step allows us to calculate the probability of the universe's transition from the "empty" state to state $S_2 = (S_2, \gamma_2 \varphi_2)$, in other words the probability of the universe arising out of nothing. In addition Hartle and Hawking make one more assumption, that Ψ_0 is the wave function of the universe in its ground state (in ordinary quantum mechanics the ground state is the state in which the system is at its lowest admissible energy). If the probability of a transition from the "empty" state to any other state has a finite, non-zero value, then, according to Hartle and Hawking, we may speak of a quantum creation of the world from nothing.

5. CRITICAL REMARKS

Hartle and Hawking's paper caused a lot of excitement. By using the mathematical formalism composed of a combination of relativistic and quantum methods, a model had been constructed for the creation of the universe out of nothing. Putting it more precisely, according to this model one could calculate the probability of the universe in a certain state arising from a state that was non-existent. However, we have to distinguish the psychological effect evoked by the comments on the Hartle-Hawking model (including its authors' comments) from the rigorous analysis of the model.

Above all we have to realise that the Hartle-Hawking model is not the cosmological application of a well-established theory of quantum gravitation, as we would like it to be, but an extremely hypothetical attempt to make a provisional model stand in for such a theory. It is a hybrid model, one that is not derived from any general laws or principles, only the result of constraining two different methods (integration over paths and the geometry of space-time) to collaborate with each other. Furthermore, the model is based on three fairly arbitrarily chosen assumptions. The assumptions are as follows:

First, the replacement of the time coordinate t by the imaginary time coordinate it. This operation allows the integration over paths to be accomplished,[9] but it is based on entirely technical grounds. Hartle and Hawking bolster this operation with the claim that thanks to it they have obtained a universe "with no boundaries", which in turn is to make the world "self-explanatory." But we must remember that in the model they propose the universe is represented not by space-time, which may or may not have a boundary, but by a wave function, and we do not really know how to interpret the presence or absence of boundaries for a wave function.[10]

Secondly, the identification of the wave function of the universe with the propagator. Admittedly, this is an ingenious step, and of key significance for the whole model. Thanks to it the model works. But at the same time we should realise that this operation is an arbitrary investment, the only justification of which would be the model's theoretical success.

Thirdly, the interpretation of the wave function $\Psi_0 = K(\varnothing; S_2, \gamma_2, \varphi_2)$ as a description of "the emergence of the universe out of nothing." While the first two assumptions were in the model's "internal mechanisms," this assumption is purely interpretative in character. What's more, it is a highly doubtful interpretation. As Gordon McCabe has remarked,[11] the symbol $\varnothing$ for the empty set in the expression $K(\varnothing; S_2, \gamma_2, \varphi_2)$ does not denote a nothingness, out of which the state of the universe $(S_2, \gamma_2, \varphi_2)$ is said to have evolved, but rather no constraints on the "initial state" of the transition to the state $(S_2, \gamma_2, \varphi_2)$. In other words the expression $K(\varnothing; S_2, \gamma_2, \varphi_2)$ describes the probability of a transition to the state $(S_2, \gamma_2, \varphi_2)$ from "anything whatsoever" rather than from nothingness.

A further reservation may be added, this time a philosophical one. Even if we agree with Hartle and Hawking that their model describes the "quantum creation" of the universe, it is not a creation "out of nothingness" in the philosophical sense of the term. The Hartle-Hawking model assumes the existence of the laws of physics, and in particular the coordinated operation of the laws of quantum physics and relativistic physics, which is very far from the concept of nothingness, in other words from the concept of the absence of anything at all.

Nonetheless the Hartle-Hawking model played an important part in the philosophical reflections on cosmology. Notwithstanding its controversial nature, it showed how far the methods of contemporary theoretical physics may go. They are capable of approaching the great metaphysical questions associated with "the beginning of existence" – seemingly to within just one small step away. Admittedly, on closer scrutiny it turns out to be just a step away from the abyss of methods and conceptual distinctions separating physics from metaphysics, but the close approach itself to questions of this kind shows their inevitability. We are speaking not only of the traditional metaphysical questions; there are also new questions, characteristic of scientific cognition, and carrying a considerable philosophical charge. These questions pertain to the boundaries of the scientific method and to the explanation of the premises at the basis of this method. Physics operates on the basis of the laws of nature. But what is the nature of these laws, and where do they come from?

PART II

~

ANTHROPIC PRINCIPLES AND OTHER UNIVERSES

Chapter 8

$\sim$

THE ANTHROPIC PRINCIPLES

1. A COMPLEX OF THE MARGIN

*F*or a long time Man has been assigning a special place in reality to himself. Already in ancient times Protagoras said that Man was the measure of all things, and later that maxim was understood in many ways, but never in a derogatory way with respect to mankind. Frequently Bishop Berkeley is accused of being of the opinion that the world existed because we perceived it: the table exists when I am looking at it; if I close my eyes the table will cease to exist. Kant claimed that how we perceived the world was more a consequence of our perception of the world than what it was really like. And the Positivists were even less modest: they said that whatever transcended our faculties of perception, which are based on experience and precise articulation, was completely meaningless.

But alongside such tendencies there was also another process going on in scholarship – the ousting of Man from his hitherto privileged place in the universe. It all started with the Copernican Revolution. Admittedly, its main ideological consequence, the degradation of the Earth to the role of an average planet in orbit around an average star – came much later, but the process launched by Copernicus certainly left its imprint on our culture's spiritual profile. The well-known historian of art, Alexandre Koyré, claimed that the cultural shock effected by the "cosmic degradation of Man" was one of the chief factors shaping the style of thinking of the fifteenth-century German mystics. On the one hand they were profoundly moved by Man's insignificance in the face of Infinity; on the other, they tried to compensate for Man's marginalisation in the universe by drawing attention to his relations with God. Koyré held that nineteenth-century German Idealist philosophy (Fichte, Schelling, and Schopenhauer) was a direct continuation of this tradition, while Hegel merely reiterated and developed in a secularised manner some of the theses

M. Heller, *Ultimate Explanations of the Universe*, DOI 10.1007/978-3-642-02103-9_8,
© Springer-Verlag Berlin Heidelberg 2009

put forward by the mystic Jakob Boehme. We may regard Positivism as a reaction to this sort of philosophy, but essentially it was a still more secularised version of the same tendency – the focusing of attention on human perception and the recognition of the boundaries of that perception as the bounds of all meaning whatsoever.

By the early twentieth century philosophers of science had reconciled themselves with the statement that "objective knowledge" was knowledge cleansed of "the human element," and that in the empirical and natural sciences there was no room for subjective factors. Nonetheless the "complex of the margin" still lingered in their cultural sub-conscious. No wonder then that when calls for a "revaluation of Man" appeared in twentieth-century scholarship, they immediately gained popularity both with the natural scientists who dabbled in philosophy as well as with the general public. The first signal of this came in the first half of the century, in as avant-garde a field as quantum mechanics in its youthful stage. The point in question was the measurement issue, a key problem for science. Prior to the taking of a measurement all that may be assigned a quantum system (e.g. an electron) is a probability that it has a certain property (e.g. a given position). The property is given an actual value only once the measurement has been done. So much the mathematical formalism of quantum mechanics; however, from this point it's only one step to the claim that it is the observer who creates the physical reality at the moment of measurement. Soon a whole spectrum of such interpretations had emerged. Not much insight is required to notice that at least some of them would espouse the character of an "ultimate explanation": if there were no human observer there would be no physical reality. Man was being promoted to the rank of the ultimate instance of explanation.

A second signal of this type appeared in the latter half of the twentieth century in cosmology, when progress made in this science had entered on a fast course, chiefly thanks to the advances in techniques of observation. It turned out that the existence of life – and all the more so of a rational observer – on at least one planet in the universe imposed some very rigorous constraints on admissible cosmological models. From here it was just one step to the statement that the universe was as it is because we were here. The diverse versions of this ideology are known as the "anthropic principles." Some of them also stake a claim to being ultimate explanations. In this chapter we shall take a closer look at this.

2. THE ERA OF MAN

It all began still in the early twentieth century. Already Eddington observed some interesting numerical relationships between the magnitudes characteristic of the world in its cosmic scale and those characteristic of its microscopic scale.

A comparison of the magnitudes characteristic for the two scales reveals that the dimensionless ratios between them are numbers of the order of 10^{40}.[1] Let's take a look at this from another angle.

Let G be the gravitational constant, ρ_0 the mean density of matter in the universe, and T the age of the universe.[2] It turns out that if we take the square of T and multiply all the other parameters by each other, we obtain a result of the order of unity,[3] viz.:

$$G\rho_0 T^2 \sim 1$$

What an amazing result! The age of the universe is growing all the time, and the universe is expanding, so its mean density is decreasing, but the product of these two magnitudes and the gravitational constant remains constant! Why are we alive precisely at an instant in time when such a relationship holds? A coincidence? Physicists don't like such coincidences. There must be some deeper reason behind it. Dirac suggested what seemed to be the obvious explanation: presumably the gravitational constant was not constant, but was slowly changing such that $G\rho_0 T^2 \sim 1$ held for all eras.[4] It turned out, however, that the rate of change in the gravitational constant required by Dirac was high enough to be observable not only in the movement of the planets, but also in the movement of the earth's crust. But we do not observe any phenomena of this kind. The mysterious equation $G\rho_0 T^2 \sim 1$ was still calling for an explanation.

Sometimes solutions can be surprisingly simple. In 1961 a brief note by R.H. Dicke was published in *Nature*, marking a breakthrough in the approach to this issue.[5] Dicke observed that the problem was not in the gravitational constant G, but rather in the age of the universe T. He pointed out that in an evolutionary universe life could not appear in any arbitrary era, but only within a certain limited interval of the universe's age. This limitation was an outcome of the physical conditions necessary for life to appear. The first of these conditions was that the universe, and therefore also the galaxy, should be old enough for chemical elements other than hydrogen to have been created in it. "It is well known that carbon is required to make a physicist," he wrote. This last sentence turned well-nigh proverbial. Indeed, carbon was a key issue. In the young universe there was only hydrogen. "The Era of Man" could have not ensued until the series of nuclear reactions within the interiors of stars had produced carbon; but neither can it come any later, when there are no more stars hot enough to provide a sufficient amount of energy to the surface of a planet endowed with life. Once Dicke had formulated these constraints in the language of the data drawn from the theory of stellar evolution it turned out that in "the

Era of Man" the age of the universe had to be related to the gravitational constant and the mean density of matter, in terms of order of magnitude, by the formula $G\rho_o T^2 \sim 1$. The relationship is by no means a coincidence; we simply could not exist in any other era, for either there would be no carbon, the building material of organic chemistry, or the universe would be too cold to support life.

3. CARTER'S LECTURE

1973 marked the fifth centenary of the birth of Copernicus. To celebrate the occasion the Cosmological Section of the International Astronomical Union held a symposium in Kraków. The subject was a comparison of the cosmological models with observational data. One of the sessions was presided over by John Archibald Wheeler. During the discussion he invited Brandon Carter to present his ideas on "Man's place in the universe." Carter asked for time to prepare, and on one of the following days delivered a fairly long lecture. Later its extended version entitled "Large Number Coincidences and the Anthropic Principle in Cosmology" was published in the proceedings of the symposium.[6] It was in Carter's lecture that the expression "the anthropic principle" appeared for the first time. Later its author was to say that if he had known that his ideas would cause such confusion he would never have made them public. We shall try to present them with the special amount of attention that they require.

Carter followed and developed the line of thinking Dicke had initiated. Summarising his ideas, we may say that he used the expression "the anthropic principle" in the sense of the mode of reasoning applied to investigate and show the relationships between certain parameters characteristic for the universe and the possibility of life emerging in it. On account of the nature of these relationships he made a distinction between the weak anthropic principle and the strong anthropic principle.

The weak anthropic principle boils down to the following statement: We are observing the universe from this particular position, in this particular era, and we see it in this particular way, since we would not be capable of life in any other place or time. In this sense, and only in this sense, may we say that our existence is the explanation of the particular properties of the universe that we observe. However, this is not a causal sense: it does not mean that our existence is the cause of these properties of the universe. On the contrary: we are the factor which is a consequence of cosmic evolution; but for mankind to be able to come into existence, cosmic evolution had to bring the universe into a particular state. No wonder, then, since we are here, that we observe the universe precisely in this state.

Formulated in this way, the weak anthropic principle is a typical selection principle. We are observing a universe with certain particular properties. Out of all the possible models of the universe we must select only those which allow for the existence of these properties (above all it is a question of the selection of time and place). We reject all the other models as incompatible with observation.[7] If there is anything amazing about the anthropic principle understood in this way it is only that it eliminates so many of the cosmological models. The existence of life, at least on one planet in the universe, turns out to be a highly restrictive condition for models of the universe.

Carter formulated the strong anthropic principle in the following way: "the universe must be such as to admit the creation of observers within it at some stage." We have to concede that such a formulation is a bit confusing, and it did indeed confuse many authors, who read it as containing a postulate of causality, or even a covert assumption of the existence of an intelligent Creator.[8] However, if we carefully read Carter's argumentation right through to the end, we get a precise exposition of his intended meaning. He draws attention to the already well-established fact that a slight perturbation in the universe's initial conditions would have led to a universe in which biological evolution could not have come about, and life would have been impossible. However, it cannot be denied that we do exist; therefore the initial conditions were such as to make this possible. Such initial conditions need not have been intentionally determined, although such an eventuality is not to be ruled out a priori. The initial conditions might have resulted from the interaction of physical laws we do not know of today, e.g. deriving from the fundamental theory we are still looking for; while the potential for biological evolution could have been only a side effect. The strong anthropic principle applies not only to the initial conditions, but also to the values of the fundamental physical constants and the other parameters characteristic of the universe (if their values had been slightly different, life would have been impossible). The time factor does not come into this line of reasoning, as in the case of initial conditions: the initial conditions were determined at a certain time, but life did not appear until much later. It was the time factor which could have led some authors astray, making them see the strong anthropic principle as entailing the attributes of causality.

The strong anthropic principle is a typical example of "reasoning back": from the consequence (we exist) to the necessary condition (the right initial conditions, the right values of the constants and other parameters). Carter did not intend the strong anthropic principle to savour of finality, which does not mean that it is not worth while considering its teleological versions as well. We shall do so below.

Carter underpinned the line of reasoning which led him to the formulation of the strong anthropic principle with the following heuristic picture. Let's imagine an ensemble of universes (in Carter's terminology) which may be described "by all conceivable combinations of initial conditions and fundamental constants." He did not ascribe a real existence to this ensemble of universes, but considered it purely as a mental construct to dramatise his line of reasoning. He wrote: "The existence of any organism describable as an observer will only be possible for certain restricted combinations of parameters, which distinguish within the world-ensemble an exceptional cognisable subset." Of course he meant "cognisable" in the sense that only a universe belonging to this subset may be cognised by an observer living in it. The drama of the situation rests in the fact that the "cognisable" subset is dramatically small. In view of this it is hard to resist the impression that our existence is something that has been "fine-tuned" – fitted very precisely into the entire structure. No wonder that the anthropic principles evoked such a tempestuous wave of discussion, in which the participants did not refrain from excursions into metaphysics. It was in this context that speculations arose on the existence of "all possible" universes not just as an attitudinal metaphor, but as a physical reality.

Chapter 9

~

NATURAL SELECTION IN THE POPULATION OF UNIVERSES

1. THE MULTIVERSE

The concept of an infinitely large number of universes, also referred to as the concept of the multiverse, grew out of the discussion concerning the anthropic principles. However, soon it started to appear in other contexts, and even developed a quantitative approach. This was true of the chaotic cosmology proposed by Linde, who constructed an inflationary cosmological model that led in a natural way to the "continual production of universes" (see Chap. 6). Although in his declarations he has often invoked philosophical motives (he wanted to secure "an eternal existence" for the family of universes), the chaotic inflation model was not prompted solely by philosophical inspiration.

One of the most radical and controversial concepts of the multiverse is the idea Lee Smolin presents in his book *The Life of the Cosmos*.[1] Smolin introduces his concept as a falsifiable cosmological model, but in the heat of discussion and his impassioned journalistic approach he has not shied away from displaying his preferred worldview.

Smolin's concept offers an excellent opportunity for a review of some of the typical problems besetting the idea of the multiverse as presented in a range of versions. This chapter will look at them.

M. Heller, *Ultimate Explanations of the Universe*, DOI 10.1007/978-3-642-02103-9_9,
© Springer-Verlag Berlin Heidelberg 2009

2. THE NATURAL SELECTION OF THE UNIVERSES

The general picture of the universe proposed by Smolin is basically no different from the picture presented in Linde's chaotic cosmology. In both models a parent universe produces descendant worlds in which the physics differs from that in the parent world. However, the mechanism by which universes are produced differs in the two models. In Linde's model inflation and quantum fluctuations are responsible for the generation of universes, whereas Smolin's version is based on two assumptions and an idiosyncratic understanding of the selection principle.

His first assumption concerns the problem of the singularities. Smolin assumes that the quantum effects of gravitation "prevent the formation of singularities, at which time starts or stops."[2] Hence, when a collapsing object – a universe or a massive star – reaches its critical density, its shrinking transforms into expansion (following a "bounce") and the whole process gives rise to a new universe. The Big Bang, too, according to Smolin, might have been the result of the collapse of another object in another world. Note that this is a very big assumption, which should be a programme for the construction of a cosmological model rather than an assumption.

Let's put Smolin's second assumption in his own words:

> The simplest hypothesis I know of is to assume that the basic forms of the laws don't change during the bounce, so that the standard model of particle physics describes the world both before and after the bounce. However, I will assume that the parameters of the standard model do change during the bounce. How do they change? In the absence of any definite information, I will postulate only that these changes are small and random.[3]

Smolin points out the resemblance of this process to the genetic inheritance of features in the living world. Descendant organisms preserve an essential resemblance to their parents, but in outcome of diverse genetic mutations may differ from them in certain respects. Such a mechanism ensures the potential for development and stimulates biological evolution. The situation is analogous as regards the evolution of the multiverse. After a long period of the generation of successive universes, the multiverse comes to be dominated by worlds which contain a large number of black holes, which generate the greatest number of descendants. And that, according to Smolin, is the essence of the selection principle in the universe population. He writes:

This is the principle we have been looking for. It says that the parameters of the standard model of elementary particle physics have the values we find them to have because these make the production of black holes much more likely than most other choices.[4]

For of course, on the grounds of the selection principle, the event with the highest probability is that the universe in which we live belongs to that most numerous subset in the multiverse which is most prolific of black holes. Hence the set of parameters characteristic of the physics of our universe should be the most favourable for the production of black holes. We don't know whether this is indeed so, but that's what Smolin has forecast.

There is yet another assumption that plays a significant role in Smolin's line of reasoning: that the same set of initial conditions, values for the physical constants and other parameters characteristic of our universe favouring the generation of black holes also enables the onset of biological evolution and its continuation up to the appearance of conscious creatures. This assumption is completely independent of the former ones, but without it Smolin's entire concept would founder. Thanks to it we obtain an explanation why the world we live in is as it is and not any otherwise. But isn't this an explanation that begs the question? And it is certainly not an "ultimate" explanation. The problem of justifying the multiverse itself has not been touched on. In particular, according to Smolin, the laws of physics are the same in every universe and the question of where they have come from is still waiting to be answered. However, Smolin hopes that perhaps one day it will be possible to explain the existence of the laws of physics on the principles of "natural selection."

3. SITUATIONAL LOGIC

It's fairly easy to notice that probability theory plays a key role in Smolin's concept. Basically the principle of selection boils down to a game of probabilities. Karl Popper had pointed out that in a large enough collection of component members competing with each other in some respect, on the strength of not much more than just probability theory, the selection principle will start to operate within the collection. In his *Intellectual Autobiography* Popper wrote, "Let there be a world, a framework of limited constancy, in which there are entities of limited variability. Then some of the entities produced by variation (those which 'fit' into the conditions of the framework) may 'survive', while others (those which clash with the conditions) may be eliminated."[5] Popper called this mechanism "situational logic" and held that it created a situation "in which the idea of trial and error-elimination, . . . becomes not merely applicable, but almost logically necessary."[6] Popper formulated this concept for the purposes of his analyses of the methodological status of Darwin's theory, and hence could regard it as "almost logically necessary."[7]

We have no difficulty in seeing that Popper's "situational logic" is also applicable to the population of universes in Smolin's concept. But with one important reservation – in the case of universes there is no environment to which they could "adapt" (unlike the situation in biological evolution). However, it turns out that "situational logic" also works in the absence of an environment. Gordon McCabe has compiled a set of precise axioms and shown that if they are met in a collection, then "natural selection" will set in almost automatically in that collection[8] (he makes no reference to Popper and does not use the term "situational logic"). His axioms (in a simplified form) are as follows: 1. The objects belonging to the system must have certain characteristics which make them differ from each other. 2. These objects must have a finite lifetime. 3. The characteristics of these objects must include some which do not change throughout the object's lifetime and which define the type to which the object belongs. 4. For every object there exists at least one object which generated it. 5. An object's characteristics are at least partly inheritable (reproducible). 6. In the process of reproduction characteristics are not copied exactly, instead mutations occur. 7. For an object of a particular type the birth rate and mean lifetime depend on its type (viz. they may differ for different types). If all of these conditions are fulfilled in any system then a selection process ensues in that collection on the grounds of the laws of statistics. Note that none of these conditions relates to an environment. In biological evolution the environment determines the generally finite set of resources indispensable for survival, and competition for access to those resources becomes an important factor of selection. But in the general case the conditions enumerated above trigger the inception of the mechanism of evolution. This is the situation pertinent to the population of universes in Smolin's concept.

4. CRITICAL REMARKS

Does Smolin's concept give a satisfactory explanation why "we live in a universe which is as it is, and no otherwise"? Unfortunately not. Above all, we should remember that Smolin's concept is based on strong assumptions and we have no guarantee – neither a theoretical one, nor an empirical one – that these assumptions hold true in reality. The arguments Smolin produces to support them effectively reduce to propaganda devices.

What's more, the existence of the multiverse itself demands an explanation. We may imagine that its explanation will prove no easier than the explanation why conditions which are life-friendly have developed in one universe.

But the existence of a multiverse is not the only point that poses difficult questions. Some of its properties are also problematic. Notice that the conditions cited above necessary for a selection mechanism to be activated make up a restrictive set of requirements. There are "infinitely more" possible systems which do not meet these conditions than there are ones which do. This applies to possible families of universes as well. Thus the question arises why the population of universes considered by Smolin belongs to that special sub-family of all possible families of the multiverses, and we belong to the sub-family of families in which the sub-family of universes postulated by Smolin exists. For if our universe belonged to a different sub-family we could not explain why there are life-friendly conditions in our universe. The superfluity of this ladder of explanations is striking and, what's more, entails a risk of regression ad infinitum.

By contrast let's note that in the family of universes referred to by Brandon Carter to illustrate the strong anthropic principle (see Chap. 8) there is no need for a selection mechanism. It is simply an ensemble of all possible universes, in which by virtue of the definition (since it is the ensemble of all the *possible* universes) there must be at least one that is life-supporting. In Smolin's concept it is not a loose ensemble or collection, but a *population*, of universes related to each other by the mechanism of selection. The anthropic principle does not impose conditions as demanding as those in Smolin's concept on the family of universes.

We shall round off this part of our examination with a conclusion as formulated by McCabe:

At best, Smolin has merely established a conditional probability: given the existence of a universe population which supports evolution by natural selection, there is a high probability that a life-permitting universe will exist. Even this conditional probability is dependent upon Smolin's postulate that the parameter values which maximise black hole production are the same parameter values which permit life.[9]

5. IS LIFE CHEAPER THAN A LOW ENTROPY?

These critical remarks may be supplemented with an observation made by Roger Penrose.[10] Biological evolution certainly requires highly specific conditions. One of them is the second law of thermodynamics, without which evolution would not work. Since, in line with this law, the universe's entropy is growing, then retrospectively it must have been less and less the further back in time. In other words, however specific the universe might be at its present stage of evolution, with life on at least one planet, it must have been even more specific at the earlier stages of its evolution, when life had not yet developed (since entropy was lower). Penrose points out, and supports this observation with a detailed calculation, that in accordance with the philosophy of the anthropic principles the selection of a universe out of all the possible universes, in which life came into existence by sheer random chance with no preceding phase as required by the second law of thermodynamics, is far more probable that the selection of a universe like ours with the second law of thermodynamics applicable in it from the very beginning.

In Penrose's opinion it would have been far "cheaper" to produce the entire Solar System along with its inhabitants by means of random collisions of particles than to explain why the universe had such a low entropy at the beginning.[11] He thinks that this fact may be explained when we have a prospective theory of quantum gravitation. That is why resorting to anthropic arguments is simply premature.

6. FALSIFICATION

Finally one more remark concerning both Smolin and many other adherents of the multiverse idea. Very often they claim that their concept is scientific because it is falsifiable. This criterion as a test of whether a hypothesis is scientific or not was put forward by Karl Popper, who said that if a hypothesis is not open to falsification by confrontation with the results of experiment or observation, then it does not qualify as scientific. Discussions are still going on in the philosophy of science as to how exactly the principle of falsification functions in science, and to what extent it discriminates between scientific and unscientific ideas. However, there is no doubt that Popper's criterion gives an accurate description of certain aspects of the practice of science, and because of this it is often invoked by natural scientists, albeit sometimes in a not very critical manner. This often happens in disputes on the multiverse idea.

So, for example, Smolin maintains that his concept is falsifiable, ergo scientific, since it predicts that the universe in which we live contains many black holes (because it belongs to the family of life-supporting universes). I doubt whether any methodologist worth his salt would admit such a prediction as a genuine falsifier of the hypothesis proposed. His main criticism would be levied against its vagueness. It is not clear what is meant by "many," and with respect to what. According to Smolin's concept a life-supporting universe should contain many black holes in comparison with other universes. How are we to determine the quantitative rules for such a comparison and, above all, how are we to carry out the comparison?

Max Tegmark, another protagonist of the multiverse idea, was even more nonchalant with respect to the criterion of falsification. In his excitement to convince the reader of the scientific nature of the idea of "parallel universes" (as he called it) he cited the following as an instance of falsification: "For instance, a theory stating that there are 666 parallel universes, all of which are devoid of oxygen makes the testable prediction that we should observe no oxygen here, and is therefore ruled out by observation."[12] So, although it is a false theory, none-theless, on the grounds of falsification, it is still a scientific theory.

We should realise, however, that not every statement (theory, model, hypothesis) the consequences of which may in principle be compared with observational or experimental results may be regarded as falsifiable, in the sense normally ascribed this concept in the philosophy of science. Taking Tegmark's style of comparison further, let's imagine that someone before the age of space flight had claimed that the other side of the moon, which is not observable from Earth, is painted red and carries the inscription, "Coke is it!" in big white letters. It would definitely have been a falsified (and therefore falsifiable) prediction, but it could never be treated as a test of whether the hypothesis was scientific or not. It's true that no theory or hypothesis which is not falsifiable even in principle may be regarded as scientific, but not all statements which are falsifiable (in the more colloquial sense of the word) may be regarded as scientific. The question of criteria distinguishing science from what is not science is a difficult methodological problem. Anyone who wants to write on this subject would do well to first look up the copious literature devoted to it.[13]

Chapter 10

⌒

THE ANTHROPIC PRINCIPLES AND
THEORIES OF EVERYTHING

1. THE SEARCH FOR UNITY

*M*odern physics was born of the union of the "terrestrial physics" and the "celestial physics." In the Aristotelian paradigm of science there was a rigid demarcation between the "sublunary world" and the "supralunary world." The former was the domain of a physics of changeability and destruction; in the latter eternal motion about circular paths were proceeding with neither loss nor expenditure. The work of Copernicus, Kepler and Galileo shook this dualism severely, and Newton's synthesis finally laid it to rest: the same force of gravity made the apple fall to the ground and held the planets in their orbits around the Sun. This search for unity was inscribed into the progress made in modern science. The next milestone on the road to unification was the work accomplished by Maxwell, who unified the electric and magnetic phenomena, hitherto known as separate, in a single theory of electromagnetism. The first overt formulation of the programme to unify the whole of physics was expressed by Albert Einstein, who said that gravitation and electromagnetism should not be treated as two independent interactions and right until the end of his life continued to search for a theory which would unify them. Unfortunately his attempt was bound to fail; as it later turned out, there are two other fundamental interactions apart from gravitation and electromagnetism: the strong nuclear (hadron) force, and the weak nuclear (lepton) force. The final unification could not be partial; it had to encompass all the interactions. The road to unification turned out much more difficult than had been supposed. Currently all we have

M. Heller, *Ultimate Explanations of the Universe*, DOI 10.1007/978-3-642-02103-9_10,
© Springer-Verlag Berlin Heidelberg 2009

achieved is the unification of the electromagnetic and the weak nuclear force, into what is known as the electroweak force. We have managed to do this thanks to the Salam-Weinberg theory. The electromagnetic and weak nuclear forces unite at energies higher than ca. 100 GeV (gigaelectron-volts); at lower energies they act as separate forces. We have managed to achieve energies of this order in the elementary particle accelerator at CERN near Geneva, and the Salam-Weinberg theory has been confirmed experimentally. Energies of this order were typical in the universe when it was 10^{-11} s old.

Extrapolating back from this success, we may infer that at energies of the order of 10^{15} GeV the electroweak and the strong nuclear interactions will unite. Energies of this order were prevalent in the universe when its age was 10^{-35} s. Obtaining such energies is at present well beyond the capacity of the laboratories we have on Earth. There are several theoretical models for the unification of the electroweak and strong nuclear forces (referred to as the Grand Unification), but so far we have no means of telling which is the right model.

On the grounds of theoretical premises we may assume that at energies of the order of 10^{19} GeV, characteristic for Planck's threshold, all the interactions including gravitation will be unified (and this is referred to as the superunification). The ultimate unification in physics will have been achieved when the same theory unites all the fundamental interactions and integrates them with a unified theory of gravitation (general theory of relativity) and quantum physics, viz. when a theory of quantum gravitation is devised. This as yet non-existent, fully integrated theory of physics is sometimes referred to as the Theory of Everything. The path taken by many of the research programmes being conducted now is oriented towards the achievement of this Theory.

The question arises whether this kind of trend to unify is not in opposition to the concept of a multiverse, or at least to some of its versions. As we recall, in many of them the birth of a new universe is attended by perturbations of the values of the fundamental physical constants and/or other parameters characteristic of the universe, or even by the modification of some of the laws of physics. Is this not a tendency towards the proliferation of variety, going in the opposite direction to unification? In this chapter we shall try to consider this question more closely.

2. CAN THE STRUCTURE OF THE UNIVERSE BE CHANGED?

At the foundation of practically all the mechanisms for the "generation of universes" there is a (tacit) assumption that if we perturb our universe's initial conditions slightly, we shall obtain a slightly different history of the universe, but one which may also develop in time. By analogy, if we perturb a parameter characteristic of our universe, we will get a universe that works, but in a somewhat different way. And finally, if we introduce a small perturbation into a physical constant or slightly modify a law of physics, we will obtain an equally good, though somewhat different set of physical laws. The formulations of these three (tacit) assumptions should, however, be regarded only as an approximation. For the universe appears to be far too integrated to allow of a distinction of its characteristic parameters from its initial conditions or the laws of physics. From the methodological point of view it's even better if the parameters characteristic of the universe are not determined a priori but are a consequence of the laws of physics, in other words are an integral part of their structure. Also the laws of physics may determine the initial conditions. As we remember, there is a tendency in cosmology for cosmological models not to require initial conditions but instead to be fully defined by the laws of physics (cf. the Hartle-Hawking model in Chap. 7). On the other hand, we cannot rule out the situation in which the initial conditions applicable to the entire universe would perform the functions of the laws of physics. So we see that drawing a distinction between the laws of physics and the initial conditions is not so obvious as it might seem at first glance. These methodological difficulties show that the universe is much more of an integrated entity structurally than our terminological conventions might suggest.

At present the only distinction that seems to be uncontroversial is that of the fundamental physical constants. They differ sufficiently clearly from the initial conditions. They are part of the laws of physics, but not in themselves the laws of physics. But even here we encounter a hint pointing to the "integrated nature" of the structure of the universe. Just as for the universe's other characteristic parameters, there is a prevalent opinion that it would be a good thing if all the physical constants could be deduced from some more fundamental principles. Attempts to do this are being undertaken. For instance, people are trying to deduce the value of the fine structure constant, the elementary electric charge, the masses of the elementary particles. These attempts seem to conflict with the practice of the creation of "other universes" by means of perturbing the various physical magnitudes. The only chance for a successful deduction of the values of the physical constants is to show that they are a consequence of the structure of the universe in such a manner that any perturbation whatsoever of that structure would eradicate the possibility of such a consequence.

The situation is similar to one we often experience in our everyday lives. If we have a fine and fairly complicated mechanism we know all too well that even a slight perturbation in a detail will make the whole mechanism stop working. And we shall have to call in an expert to set it right.

3. RIGID STRUCTURES

The exposition of these arguments does not, of course, disprove the possibility of the existence of many universes, nonetheless it does show that we should proceed very carefully. Let's make our discussion somewhat more rigorous.

Let S be a mathematical structure determined by parameters $p_1, p_2, p_3 \ldots$ Let us assume that there is a mathematician who wants to generalise the structure of S by manipulating with some of the parameters. There is a method in mathematics which may be used to do this. It involves the *deformation* of the structure by a change in the value of one or more of the parameters, which leads to a generalisation of the structure. The term "deformation" denotes a certain technique the details of which we shall not go into. We shall only give two informal examples to illustrate just a few of the aspects of deformation. As we know, classical mechanics has a well-defined mathematical structure with two physical constants: Planck's constant, which takes the value of zero; and the speed of light, which in classical mechanics takes the value of infinity. We may deform the structure of classical mechanics by changing the value of Planck's constant, ascribing a non-zero, positive value to it. Thereby we obtain the structure of quantum mechanics. Or we may deform the structure of classical mechanics by ascribing a positive, finite value to the speed of light, and thereby obtain the structure of the special theory of relativity. It need not be added that this role played by the two constants in question did not come to light until the appearance of quantum mechanics and the special theory of relativity. What we are concerned with here, however, are not the historical facts but the properties of this method.

Deforming the structure of a theory of physics is, then, in a certain sense, the opposite of a method which is well-known and much discussed in the philosophy of science – the *reduction* of a new theory to an earlier theory. For example, quantum mechanics reduces to classical mechanics as Planck's constant tends to zero; and the special theory of relativity reduces to classical mechanics as the speed of light tends to infinity. In such cases we speak of a principle of *correspondence* between the new theory and the earlier theory. By applying this principle it may be shown that the development of physics is not random: a new theory does not invalidate the theory in use hitherto but merely "absorbs" it as its limiting case. On the other hand, by applying the deformation method we may predict the direction new developments in physics will take: if we deform the structure of the current theory we will be able to anticipate the structure of the forthcoming theory. However, the snag is that deformation does not, in general, yield an unambiguous result: the same structure may be deformed with respect to various parameters and in various ways.

However, there exist some structures which are particularly "permanent." Let's consider a structure S and try to deform it with respect to a parameter p. It may happen that any change at all in p will produce the same, undeformed structure S. Such a structure is called a *rigid* structure with respect to parameter p.[1]

When we consider the prospective ultimate theory of physics, the Theory of Everything as some like to call it, we want it to have a rigid mathematical structure with respect to all the parameters. Indeed, as some authors stress, the Theory of Everything should be necessary – that is, the only possible theory. The only reasonable constraint on this rather general postulate would seem to be the requirement of rigidity in the structure of such a theory with respect to all of its parameters. Such a theory would indeed be the "only possibility" in its class. Is this requirement realistic? Can it be accomplished? It remains to be seen.

4. IMAGINATION AND RATIONALISM

In the light of the above reflections we need to investigate two possibilities. According to the first, the perturbation of any component of the structure of the universe gives rise to the destruction of the entire structure. This would indeed hold if the structure of the universe were rigid in every respect (with respect to all of its parameters). In that case the universe in which we live would be the only possible universe and there would be no chance of the production of new universes on the disturbance of its structure. And the "anthropic coincidences" would be encoded in the initial conditions of our universe, since presumably the structure of the entire universe would have required this. Any other initial conditions would break up the structure. Perhaps one day (soon?), when we succeed in developing a Theory of Everything, we shall be able to observe that they are not "coincidental," but necessary elements of the entire structure.

The second possibility is that a disturbance (perhaps only a perturbation that was not large, but what does "not large" mean?) of the structure of the universe would not bring about its utter destruction, but merely cause a variety of "adaptations" within it. We may imagine such a structure as logically coherent to the highest degree possible, but within that level of coherence there would be room for the manipulation of some of its parameters; for instance, in such a way that a (slight) change in one parameter would have to be "compensated for" by a change in some other parameter. It might be that a certain structural change would preclude the possibility of life of the kind that exists in our universe, but would not rule out the possibility of life operating on different principles. For example, in their monograph on the anthropic principle John Barrow and Frank

Tipler cite work which shows that the solutions to Schrödinger's equation (supplemented with certain analyses based on the special theory of relativity) rule out the existence of stable atomic orbits for spaces of more than three dimensions. Since stable atoms are indispensable for "the chemistry of life" Barrow and Tipler see the fact that space in our universe is three-dimensional as an "anthropic coincidence."[2] This conclusion is based on the strong assumption that a change in the dimensionality of space would not affect "the rest of physics." But of course, in line with our hypothesis of a "maximally coherent" structure of the universe, it could be that in a space with a different number of dimensions some other equation could perform the functions presently performed by Schrödinger's equation, as a result of which we would have different stable atoms, a different biochemistry, and life based on different principles.

This is borne out by the detailed research carried out by Gordon McCabe.[3] He has examined the standard model of elementary particles and shown that disturbances of the geometrical structure of space-time (its signature or number of dimensions) do not lead to the destruction of the standard model for free elementary particles, but produce different sets of particles. If we assume that universes with a disturbed space-time structure are not just a useful mental tool for the mathematics but describe other universes that really exist, then different sets of elementary particles may occur in such universes. We have to bear in mind, however, that McCabe's analysis concerns only the standard model for particles and does not say anything about the prospective Theory of Everything on this issue.

5. OUR ANTHROPOCENTRISM?

Are we not perhaps being too anthropocentric ourselves in our discussion on the anthropic principles? Can we not imagine a different biology than the one which has grown up on our planet? Perhaps we do not want to admit the idea (or maybe it just doesn't occur to us) that there might exist an atomic physics governed by an equation other than the one discovered by Schrödinger? And maybe that is why the multiverses we are conjuring up are too close to our own measure?

Such deliberations are, of course, highly speculative, but their aim is to throw light upon the equally speculative notion of the existence of "parallel worlds."

Nevertheless I would not like to go to the other extreme and deny the multiverse idea any cognitive value. Strictly speaking, the concept does not meet the criteria of science, since by very definition we have no experimental control whatsoever over other worlds, and the chances of indirectly testing their existence are negligible and based on probabilistic speculation. But the history of science teaches us that the hard core of science has always been enveloped by a ring of more or less philosophical speculations which have often played a heuristic role, suggesting valuable ideas and encouraging progress in previously unexpected directions. Today the concept of the multiverse is performing a similar role, which will be fulfilled to an even higher degree if the creative imagination is combined with rational criticism.

Chapter 11

THE METAPHYSICS OF THE
ANTHROPIC PRINCIPLES

1. THREE PHILOSOPHICAL ATTITUDES

*T*he anthropic principles force you to think. And they do so with a tremendous power of persuasion. The extremely precise fit of human existence into the structure of the universe is surprising at first, but after a while suggests various responses. The first thing that comes to mind is the principle of purposefulness. How did the initial conditions know how to fine-tune in order to make our existence possible? With such a high level of fine-tuning, the probability of a random occurrence seems negligible (in addition, for some people the idea that they exist by random chance is repugnant). Purposefulness is associated with the teleological argument for the existence of God. And indeed, the concept of the Grand Designer soon cropped up in discussions on the anthropic principles. But the Positivist and Empiricist traditions in the philosophical environment attending science were still too strong for that concept not to come up against a firm confutation. If there exists an infinite (or at least sufficiently large) number of universes fulfilling all the possible combinations of initial conditions, physical constants and other cosmological parameters, then it is no wonder that we live in a very special, life-friendly universe, since we could not have come into existence in any other universe. Therefore there is no random occurrence and no need to assume a Designer.

There is another option: simply to accept the message of the anthropic principles. And even make it as sharply defined as possible: Man did not come into existence because the universe was as it was at its origin, but the universe is as it is because Man exists in it.

M. Heller, *Ultimate Explanations of the Universe*, DOI 10.1007/978-3-642-02103-9_11,
© Springer-Verlag Berlin Heidelberg 2009

Thus we have three philosophical attitudes to the collection of issues posited by the anthropic principles. They may be summed up by keywords: Designer, Multiverse, Man. We shall look at these three groups of issues, but in a different order. We shall start by presenting two views (Wheeler's and Hawking's) on the Man – Multiverse coupling, and examining the problem of human individuality in the context of Man's potential existence in multiple copies in an infinite multiverse. Next we shall make a few observations on the subject of "Designer or Multiverse?" (though a fuller treatment of this will come in Part III), and only then, in the next chapter, will we attempt an evaluative review of the entire collection of ideas on the multiverse.

2. THE "PARTICIPATORY UNIVERSE"

John Archibald Wheeler was well-known for his "crazy ideas." Some of them developed into worthwhile research programmes and brought him fame. Whenever any of his ideas led up to a dead end, he would abandon it and think up a new one. Wheeler was not afraid of attacking the most difficult problems. For a long time he'd been looking for physical processes which could . . . call the universe into existence. But of course all processes are part of the universe, so they would have had to call themselves into existence. The wheel must come full circle. An opportunity opened with quantum mechanics, or rather its Copenhagen interpretation. According to this interpretation, certain properties of quantum objects (e.g. their spin, location, momentum. . .) do not exist until they are measured. It is the act of measurement or observation that calls these properties into being. Prior to the act of measurement a quantum object (e.g. an electron) can only be assigned a probability of being in a certain location, or of having a certain momentum. The object only acquires a given location or momentum once the measurement is carried out. Could it not be a similar case with the existence of the universe? Perhaps all the properties of the objects in the universe, together with the very existence of the universe, are only called into being by acts of observation carried out at a particular moment in the history of the universe. In other words – the universe did not have the property of existence until it was observed by a conscious observer.[1]

Wheeler was well aware of the tremendously speculative nature of his assumption, but that did not stop him from taking the idea further. A single act of observation (measurement) was required to call a particular value of an electron's momentum into existence; but perhaps to call the universe into existence required observation by all the conscious beings in the universe. Wheeler called this kind of universe a "participatory universe." Presumably this should be understood to mean that Man and the universe do not exist independently of each other, but mutually participate in each other's existence and mutually condition each other.

Unlike many other ideas generated in the discussions on the anthropic principles, Wheeler's speculations did not give rise to any new research models or projects. They simply remained as speculations not very closely connected with cosmology. Nonetheless they show how far ambitions to discover the ultimate explanation of the universe can go.

3. CREATING OUR OWN HISTORY

However, the idea of self-creation of the observer – universe system has recently re-emerged quite out of the blue in discussions on cosmology. It happened thanks to Stephen Hawking with no overt reference to Wheeler's earlier ideas.[2] Hawking, in collaboration with Thomas Hertog, simply returned to his old quantum model of the creation of the universe out of nothing (see Chap. 7). As we recall: Feynman's method of integration over paths allows us to calculate the probability of the transition of a quantum system from state S_1 to state S_2, by computing the action integral over all the possible paths between these two states. Hartle and Hawking brought this method into cosmology, replacing the states of the quantum system with states of the universe, and all the possible paths joining those states with all the possible histories (geometries) from the S_1 state of the universe to the S_2 state of the universe. By performing a clever move they eliminated state S_1 (for details see Chap. 7) and calculated the probability of the universe emerging into state S_2 out of nothing (if there is no state S_1).

Now let S_2 be the state of the universe we are now observing by means of our latest telescopes and space stations. "In the beginning" the universe was a quantum object; therefore there exists an infinite number of histories joining that "beginning" with its current state S_2. In accordance with the method of integration over the histories adapted for cosmology and interpreted in the spirit of the Copenhagen interpretation of quantum mechanics, it is the act of observation that defines the history which has led up to S_2, the currently observed state of the universe. And that very same act of observation has made all the other histories "extinguish themselves."

In Hawking's opinion, it is not a case of the history of the universe which began 13.7 billion years ago leading up to our existence; instead, the observations we are conducting now are selecting our history. The universe and the observer (mankind) are in a "causal loop." Hawking reassures us: this kind of violation of ordinary causality occurs only when we are looking at the universe from without, from "the divine perspective." But from our own perspective, the viewpoint of creatures immersed in the universe, it all looks exactly like what we are used to: we see the current state of the universe and think that it is the outcome of a single history which has occurred and which we are able to reconstruct.

It's certainly a fascinating vision, but we should look at it from a critical distance. It is based essentially on the Hartle-Hawking model, and all the critical remarks and objections raised against that model at the end of Chap. 7 apply to this idea as well. Moreover, Hawking's latest hypothesis is a substantial generalisation of his earlier model, which makes it even more vulnerable to criticism: a generalisation of a shaky hypothesis tends to be even shakier.

We should also realise that Hawking's idea queries the entirety of the scientific method as practised hitherto. The fundamental canon of the scientific method is that observational results are to determine whether a theory is sound or not; whereas according to Hawking's concept the theory (history of the universe) is adapted to the results of observation. "That, after all, denies us the chance to see if the theory matches up with observations."[3]

4. HOW MANY COPIES OF HIMSELF DOES THE READER HAVE?

We have already said a lot about the multiverse in the previous chapters, and in this chapter, which is on metaphysical speculations concerning the anthropic principles, we cannot avoid the subject, either. All the more as an exceptional number of exceptionally fantastic ideas have sprung up around the notion of the multiverse. We shall take stock of some of them – not because they deserve special attention, but because they offer a good opportunity for a review of some of the philosophical aspects of the issue, which many authors treat rather perfunctorily.

We shall skip Tegmark's idea put forward in his hypothesis that all conceivable mathematical structures not only exist objectively but are also embodied as separate, physical universes.[4] This notion, formulated in an extremely vague manner, may at most serve as an inspiration for science fiction novelists. But in the same article Tegmark expresses a statement which seems well-nigh obvious to many other authors. Tegmark presents it to his reader in a dramatised way:

> *Is there a copy of you reading this article? A person who is not you but who lives on a planet called Earth, with misty mountains, fertile fields and sprawling cities, in a solar system with eight other planets? The life of this person has been identical to yours in every respect. But perhaps he or she now decides to put down*

this article without finishing it, while you read on. The idea of such an alter ego seems strange and implausible, but it looks as if we will just have to live with it, because it is supported by astronomical observation.[5]

Tegmark's argument is simple and seems convincing. If the space of our universe extends out to infinity, which astronomical observations do not rule out, then any arbitrary configuration of atoms, even the least probable one, occurs an arbitrary number of times in it.[6] As he writes, "everything that could in principle have happened here did in fact happen somewhere else."[7]

We must admit that science, and especially modern physics, has often taught us a lesson about trusting too much to what is regarded as common sense, but it has also schooled us in sound criticism, indispensable especially when we are dealing with a concept as "wild" as infinity. In such situations it is good to consult the lesson of history. For a long time philosophers have been discussing the problem of the principle of individuation – the philosophical version of the question which sometimes even small children ask themselves: "Why am I?" An individual, especially a person with his/her own awareness, is more than just an ordinary aggregate of particles. What is it that makes something a particular, specific individual? And if a "principle of individuation" exists, does it not preclude the individual's blueprinting in many copies? The fact that even in mathematics we come across a certain "principle of individuation" suggests we should take such a possibility into account. As we all know, the set of real numbers comprises an uncountable number of elements. But in spite of this every real number occurs just once in that endless set. No number comes a second time. What's more, we are able to attribute a name to every real number (to any degree of accuracy), e.g. 0.123345..., and we can do this because each of them has its own individuality. What gives it an individuality is that every real number has properties which are proper only to itself (e.g. every real number has its own prime factorisation), and properties resulting from the order on the real number axis. This is a highly relevant example which should teach us to be extremely wary in embarking on intuitive speculations on the subject of infinity.

5. A FALSE ALTERNATIVE

Many authors have certainly received the idea of the multiverse as an alternative to the concept of a Divine Designer: one had either to accept the existence of a Rational Creator, Who had so ordered the universe that we might come into existence within it; or acknowledge that there exists a vast (perhaps infinite) set of "all possible" universes, and we live in this particular one, since we could not have come into existence in any other. The choice is to a large extent a question of one's philosophical preferences. Martin Rees considers the multiverse idea more rational,[8] while John Leslie sees the God hypothesis as more economical.[9]

In the context of this alternative it is worthwhile quoting Sherrilyn Roush, who writes:

> However, forced to choose between the God-hypothesis and the SAP [Strong Anthropic Principle] many-world hypothesis, it seems to me that the God-hypothesis would be preferable on grounds of simplicity. For notice that a many-worlds hypothesis that is an articulation of the SAP will rest its explanatory case on a claim that the physical possibility of intelligent observers was a goal for the sake of which all those worlds, including ours, became factual.[10]

The point is, however, that it is not an exclusive, "either-or-but-not-both" alternative. The existence of even an infinite number of universes does not rule out the existence of God. As someone has pertinently observed, if God is infinite, then perhaps He is not interested in anything that is less than infinity. Theologians have for a very long time been speaking of the infinite fecundity of God in His activity of creation. Furthermore, the same questions pertain to an infinite number of universes as are asked with respect to just one universe: for instance, Leibniz's question, "Why does something exist, rather than nothing?" (see Chap. 21). The philosophy of the multiverse does not remove the issue of God, but merely extends the range of the problem.

In the above reflections I have tacitly assumed the Christian concept of God (most of the thinkers I have discussed made the same assumption). But it is not a necessary assumption. In the history of human thought there have been many versions of Transcendence or the First Cause. Alongside the Theistic opinions, in which God is regarded as a Person, there have also been attitudes recognising an Ultimate Mind; a Deist, not necessarily personalised First Cause; and a Pantheist Substance. All these "keywords" are broad enough to include widely different opinions. The Ultimate Mind may assume a Christian version, not far from the ideas of St. Augustine and Leibniz. Deism has been associated with the idea of a Personal God Who created the world but entrusted its governance to the laws of nature, Himself as it were stepping aside; but Deism may also be viewed as connected with the existence of an impersonal Power permeating all the laws of nature. There is a continuum of such notions into Pantheism, which again may assume a variety of forms: from a simple identification of God with Nature, right up to the claim that the universe is the "body" of God. The notion of the universe existing within God, but God being transcendent with respect to the universe is referred to as Panentheism and is accommodated within Christian orthodoxy. We should also mention Emanationism, which over the centuries has proved a rather strong temptation for thinkers from the Christian milieu. The Church's official pronouncements have often censured this tendency, which claims that the world came into existence not by being created, but by emanation, that is splitting off from the Divine Being. We could say that the contemporary pantheistic trends are an extension of the tendency to "divinise" the world.

The following criticism is often put, also in discussions on the multiverse, against the idea of God, whether personal or not. Whenever we ask for the cause of the existence of something, we are doomed to a chain of successive explanations: B is the cause of A, C is the cause of B . . . etc. Why should we have to stop the sequence of causal explanations at God; but if we ask for the cause of the existence of God we run the risk of *recursus ad infinitum*, an infinite sequence of successive questions.

I think this is an informative criticism. It errs by reducing God to the category of other, *finite* causes. But by very definition, God is the Necessary, Self-Existing Being. The reason for His existence is within Himself. Of course one may not accept the existence of such a Being, but if one acknowledges the existence of God, then that is what His Being is. The criticism is informative because it shows that only an infinite sequence of finite causes may replace the notion of God. There are good reasons for the claim made in the philosophy of God that He is the Infinite Being. The Infinite Cause is not the first link in the chain of causes. It justifies the existence of the entire chain, even if the chain were infinitely long.

Chapter 12

~

TEGMARK'S EMBARRASSMENT

1. OTHER UNIVERSES IN PHILOSOPHY AND MATHEMATICAL PHYSICS

*T*he reader who has got this far will not need much persuading that there is a great deal of confusion, both conceptual and methodological, in the collection of issues surrounding the slogan "multiverse." Various authors understand the expression "other universes" in various ways; they employ various justifications for the need for multiverse studies and differ in their assessment of the status thereof. Some regard them as creations of science fiction, others see them as metaphysical hypotheses, and still others consider them falsifiable models deserving a place in official science.

Other worlds (universes) have been a subject for discussion in science and philosophy for a long time now, not always merely as fictional possibilities. When Leibniz was pondering on the question of God's reasons in choosing our universe as the "winning candidate for implementation" out of the infinite number of other universes (see Chap. 18), he did not treat these other contenders as purely fictional, but rather as real candidates for being created. Today philosophers continue to follow Leibniz's method, albeit in a secularised version, which they often refer to as the method of "counterfactual conditionals." By "counterfactual conditional" they mean "statements asserting that something happens under certain conditions, which are presupposed not to be satisfied in reality."[1] For instance, we wish to find out whether a certain statement is "necessary" in a given system of beliefs. To verify this, we negate the statement or modify it in some way. Then we insert the negated or modified statement into the system and examine its logical coherence. And we say we have "constructed a different

M. Heller, *Ultimate Explanations of the Universe*, DOI 10.1007/978-3-642-02103-9_12,
© Springer-Verlag Berlin Heidelberg 2009

universe." Basically every natural law may be interpreted as a counterfactual statement. For example, the law of inertia says that if there is no force acting on a body, it will move at a constant velocity and in a straight line. In our universe such a situation never happens de facto, so we call into existence "another universe" in which this law is strictly obeyed. But we only do this to learn something about our own universe. In this context the "other universe" is simply a logically coherent *description* of a certain reality. Nonetheless, we have to admit that there exist authors who insist that "other universes" are not only a description of a certain reality, but actually make up a certain reality.[2]

The concept of a set or family of worlds (universes) has also been current for a long time in the empirical sciences, quite independently of the contemporary discussion on the multiverse. In mathematics and theoretical physics studies are frequently conducted of the space of solutions to some differential equation; if it is agreed that each solution to the equation describes a possible universe (and this is done fairly often), then the space of the solutions is simply a family of universes. This kind of space may be studied using very sophisticated mathematical methods. The subject of study may be not only the particular solutions, but also the entire structure of the space: its stability, sensitivity to perturbations, the distribution of a variety of properties in the space of the solutions. Such procedures are very important in applications in physics, where the measurements obtained experimentally are never exact but always subject to experimental error. The physical situation is never modelled by a single solution to a differential equation, but by a sub-family of solutions which are close to each other and the measurable parameters of which are contained within the admissible margin of error. Moreover, the sub-area of the solution space to which a given solution belongs must have the property of stability, viz. a small perturbation in the initial conditions for a given solution should yield a solution which does not differ much from the given solution. Otherwise the margin of error would encompass radically different solutions for the modelling of a given process, in other words the experiment would confirm many very different models (within the margin of experimental error), and no empirical predictions would be possible.

This procedure also works in cosmology, and here it becomes even more like the idea of the multiverse. Cosmological models are simply solutions of Einstein's equations, with appropriate initial or boundary conditions. The space of all the solutions is often called the ensemble of universes, and is the subject of intensive study. But in view of its complexity it also poses a formidable challenge to the theoreticians. Only a narrow part of its sub-regions has been fully examined.[3]

The universes understood in this sense (viz. elements in the space of solutions to Einstein's equations) are attributed an existence of the same kind as other mathematical entities. Thus the Platonists in the philosophy of mathematics will claim that these universes/solutions exist in an abstract (but real) world of the "Platonic ideas"; while the Constructivists will say that they exist only as our constructs. Until recently no-one ascribed a physical existence to them. Curiously, the proponents of contemporary speculation on the multiverse are not very keen on calling the space of solutions to Einstein's equations an ensemble of universes. Solving Einstein's equations calls for very precise mathematical methods, while the idea of the multiverse has proved notoriously evasive of mathematical precision. Though this does not mean that there have been no attempts to introduce mathematical rigour into these issues.

2. DOMAINS AND UNIVERSES

An attempt of this kind was undertaken by George Ellis, though it brought a negative result for the multiverse concept.[4] First we have to realise that a different methodological status should be ascribed to different concepts of the multiverse. In general we have to distinguish between two different concepts of the multiverse. According to the first the multiverse is a set of domains within the same space-time, separated off from each other in such a way as to be incapable of influencing each other causally. From within our own domain we can have no observational access to other domains. Although the particular domains are de facto parts of the same space-time, many authors refer to them as "different universes." Diverse domains may be related genetically, for instance by deriving from the same domain. This is the case with Linde's chaotic inflation model, in which various universes "sprout" from other universes (see in Chap. 6 Sect. 2).

In the second concept the elements of the multiverse are genuinely separate universes. No contact at all – either causal or observational – is possible between them, and their space-times (if other universes have space-times) are completely separate.

Naturally the methodological status of the domain universes and the genuinely other universes is completely different. Other domain universes, albeit observationally inaccessible to us, may be part of the same cosmological model in the generally accepted sense of the term (as in Linde's model). Many cosmologists of the empirical and observational orientation have their misgivings about this; while others stress that not all the aspects of even well-grounded theories of physics are subject to direct observation (suffice it to mention the structure of quantum mechanics). Importantly, there may be a justified physical motive for the postulate of the existence of domain universes.

The situation is completely different for the genuinely separate universes. It is hardly imaginable that anyone will come up with hard physical evidence for their existence. Not surprisingly, the disciples of this trend invoke a range of philosophical arguments differing in their persuasiveness – from diverse anthropic motivations to the "principle of fecundity" put forward by R. Nozick and others, according to which "all that is possible actually exists."[5]

Neither concepts of domain universes nor speculations on genuinely separate universes may be directly falsifiable (see in Chap. 9 Sect. 6), but concepts of domain universes may be disproved if they are part of a cosmological model (in the standard sense of the expression) and the model itself is falsified. For instance, Linde's idea of universes continually generated in a process of general inflation would have to be rejected if it turned out that there never was an inflationary period in the history of the universe.

3. JUGGLING ABOUT WITH PROBABILITIES

It would be hard to think of a line of reasoning connected with the multiverse idea not referring, directly or indirectly, to probability theory. What is the probability of drawing a universe with initial conditions like ours out of the entire pool of universes? If there exist universes with all the possible combinations of initial conditions, then no wonder that our world belongs to the "very low-probability" sub-set of "life-friendly" ones (we could not live in any other). Etc., etc. The fact that this recourse to probability is the raison d'être of the multiverse compels us to take a closer look at the concept of probability.

In mathematics the concept of probability is a special case of the concept of measure, and probability theory a special case of the mathematical theory of measure. In the most general sense (for details see in Chap. 20 Sect. 3), measure is a function which assigns numbers to the objects measured (their "measures"). For instance, if we say that this block has a volume of 1 L, we are assigning the number one (in a defined unit) to it. If the numbers assigned to an object have the property of lying within the range from zero to one (with zero and one included in the range), the object measured is called an event, and the measure assigned it is a measure of its probability, or probability for short.

So much the mathematical definition, but why is there such good agreement between probability theory and what happens in the world? Because we are the ones who, on the basis of a long series of experiments and experiences, decide what numbers (measures of probability) are to be assigned to what events. The fact that in a long series of throws of a true die one-sixth of the throws gives a six is neither a "metaphysical necessity" nor the outcome of a mathematical law, but the result of our very long "experience of the world." It is simply one of the world's properties. The rule we lay down for the assignment of numbers (measures of probability) to particular events is called the probability distribution. If no such rule has been established, then the concept of probability is meaningless.

As soon as we apply these basic rules of probability theory to the multiverse we are faced with two salient questions: first, are we at all entitled to apply probability theory to the multiverse? And if so, does a measure of probability exist on the multiverse (space of universes)? If and only if the answer to both of these questions is in the affirmative will we have the right to consider how to determine that measure.

The former question is philosophical in character. Naturally we are not able to determine a probability distribution function on the multiverse on the grounds of experiment. What remains are philosophical motifs like a sense of simplicity, mathematical elegance, resemblance to or analogy with our own universe. They're not very objective grounds. What's more, they take for granted that probability theory is a kind of meta-law governing the multiverse. Such an assumption is justified with respect to our own universe, on the grounds of long experience, but this advantage is inapplicable to the multiverse.

The latter question is technical in character. The question of whether or not there exists a measure of probability on a given space is by no means trivial. In mathematics spaces on which there is no measure of probability do occur, and are not rare exceptions. How this relates to the multiverse will depend on what we mean by "multiverse." If it encompasses all possible universes, then there is no chance of assigning any kind of meaning to the concept of a probability measure on such a set.[6] Even if we decide to rigorously restrict the concept of the multiverse, for most cases discussed by various authors there will still be no probabilistic measure at all. We should acknowledge the comment made by Max Tegmark, a great enthusiast of the multiverse idea, as very reserved. He has written:

> As multiverse theories gain credence, the sticky issue of how to compute probabilities in physics is growing from a minor nuisance into a major embarrassment.[7]

George Ellis has attempted to present the problem of measure in the multiverse in a more rigorous way.[8] First he proposes a space of possibilities M be defined, comprising all the universes regarded as possible. All the states in which each of these universes may exist make up a space of states S. Each universe would be characterised by a set of parameters which should be treated as coordinates in the space S. To define the probability problem correctly, we would have to know all the parameters for each of the universes, along with the ranges in which they may take values. We would have to resolve the tricky problem of how to identify the same universe defined by various arrangements of parameters.

Ellis distinguishes several classes of these parameters: (1) physical parameters characteristic of the laws of physics, physical constants, properties of elementary particles etc.; (2) cosmological parameters characteristic of the geometry of each universe's space-time and material content; (3) parameters determining the possibility of the emergence of complex structures, including life and consciousness (the last two of which we do not fully know even with reference to our own universe).

Only once we have constructed a space of possibilities M in this manner may we undertake an attempt to define a measure of probability. Here again, a series of technical snags lies in wait. But let's assume that we have surmounted them, that we have a correctly defined space M and a definition of the measure of probability on it. Then, according to Ellis, we still have two unanswered problems:

First, what determines space M? What (and on what grounds) do we allow as the possibilities which have to be taken into consideration?

Secondly, what determines the measure of probability on space M? Is there a meta-law which determines what probabilities are to be ascribed to what possibilities?

These are fundamental questions. They show that when we speak of a multiverse we cannot pass over in silence the existence of meta-laws governing that multiverse, in other words the meta-physics of the multiverse. If we do not adopt such meta-laws, the answers to the above questions will have to remain absolutely arbitrary. A set of all possible outcomes with no laws or meta-laws limiting them is "mathematically untreatable." However, since the question of meta-laws lies in the sphere of pure conjecture, we should be speaking not of the meta-physics, but simply of the metaphysics of the multiverse.

Finally we should note that the construction of space M put forward by Ellis is purely postulative in character. It could be done only for a very limited class of universes. But from the point of view of the purposes for which the multiverse ideology has been developed, such a sub-class would be extremely unrewarding (in physics drastically simplified models of this kind are called "toy models"). No wonder Tegmark feels "embarrassed" when considering probability with respect to the multiverse.

4. AN APOLOGY FOR THE MULTIVERSE

After this unsparing criticism of the multiverse idea, I would like to refer back to the remarks I made at the end of Chap. 10. I expressed an opinion that in every age the hard core of science is surrounded with a broad band of speculation. Some of these speculations play the role of an inspiration, or may play such a role in the future; others develop a more rigorous form and become genuine issues in science. But the ring of speculation also holds numerous ideas which are totally useless and will be remembered only by the more erudite historians of science. Perhaps the band of speculation surrounding the hard core of science is today broader than ever before. This appears to be an outcome of the tremendous progress made in science, which has conquered (almost) all the territories there were to conquer (though isn't that what scientists have believed in all ages?). Today the band of speculation must be really broad, since it holds so many diverse concepts of the multiverse. The fact that they are being hotly debated by distinguished scientists shows that the idea of the multiverse is beginning to play the role of an inspiration for science. Occasionally voices may be heard that these discussions are a forecast of an evolution in scientific method heading in the direction of a relaxation of its rigours and an acknowledgement of the right of directly unfalsifiable hypotheses to a place within science. However, I would be extremely wary of pressing such claims. It cannot be ruled out that when at last we have a Final Theory, it will once again change the global perspective on science. The current difficulties in arriving at a Final Theory certainly do not have to mean that there is no such theory; and the multiverse hypotheses will have played their heuristic part if they bring us onto the right trail leading up to it.

PART III

~

CREATION OF THE UNIVERSE

Chapter 13

~

THE DRIVE TO UNDERSTAND

*B*oth in contemporary cosmological research as well as in the various aspects of speculation continuously arising in connection with it there is a persistent urge to tackle more and more radical questions. In Chap. 1 I called this tendency "the longing for ultimate explanations." I have tried to trace its progress in the recent developments in cosmology. Admittedly, our account of the theories, models and the more speculative reflections has been far from a full overview, but I hope it has presented a sample representative enough to serve as a basis for at least some schematic conclusions.

First of all, it is rather obvious that what is at issue is, generally speaking, a justification of the universe: its existence, its laws, the way it works. But insurmountable problems start as soon as it comes to taking a closer look at that "justification." Perhaps this is so because, as the history of science has often shown, putting the right question is only possible once we know the right answer. But so far no answer has emerged to the question of an ultimate explanation of the universe, and there does not seem to be much of a chance for one emerging "within a finite period of time." Nonetheless, the review we have carried out enables us to observe certain regularities both as regards the asking of questions as well as the searching for answers to them.

There is certainly nothing novel about the statement that very often the endeavour to understand something boils down to breaking down that thing into its parts and trying to reduce it to its most fundamental components. This method has for a long time been the prime path for progress in science, and it is to this procedure that physics, both in its classical version as well as in its most modern embodiments, owes its biggest successes. This approach is distinctly present in contemporary cosmology, and is becoming even more prominent.

M. Heller, *Ultimate Explanations of the Universe*, DOI 10.1007/978-3-642-02103-9_13,
© Springer-Verlag Berlin Heidelberg 2009

Although relativistic cosmology started with the search for solutions to Einstein's equations which would come up with models of the universe on its grandest scale, from the very beginning questions concerning the processes which may have occurred in space-time were lurking in the background. The status of cosmology became firmly established only once its global geometrical methods merged with local physics. This coalescence soon led to the emergence of the standard cosmological model, and today the successes and problems attendant on the standard model are staking out cosmology's paths of development. One of these directions is "inward bound." It is no coincidence that physicists working on elementary particle research and wanting to test their ideas look to models of the early universe, where they have not only high enough energies but also the chance to make empirical predictions that may be verified by contemporary astronomical observation. The work to construct a quantum cosmology go even deeper, to where the foundations of cosmology meet and unite with the laws of physics.

However, cosmology has never abandoned its original perspective – the perspective of wholeness. In Part II of this book we saw that in recent times the scope of speculation is expanding in this aspect of the research, too – no longer is there talk just of one universe, but of an entire family, perhaps an infinite family, of universes. But even if we take a sceptical attitude of this idea, in cosmology we shall still have to consider an infinite number of possible universe, if only because there exists an infinite number of solutions to Einstein's equations which may be treated as possible universes, many of which are interesting from the theoretical point of view.

Reductionist methods and holistic (integrative) methods have been known and practised for a long time in science, but the efforts made in cosmology to justify the universe have given rise to a new phenomenon – an idiosyncratic linking up (or indeed identification) of these two trends. Certainly the discovery of a superdense state (the Big Bang) at the beginning of cosmic evolution is a success which must be attributed to global methods. The cosmological singularities made their first appearance in solutions to Einstein's equations, which were to describe the global structure of the universe. Furthermore, thanks to the global methods it has been shown (in theorems on the singularities put forward by Hawking, Penrose, and others) that there are no simple means to eliminate the singularities from cosmology. But the really exciting things started to appear when the methods of high-energy physics were applied to the reconstruction of the physical processes that must have occurred in the neighbourhood of the initial singularity. And moving down even further, down to Planck's threshold and beyond, the global becomes indistinguishable from the local. The difference between what is global and what local becomes blurred and finally disappears altogether. Even our idea of the universe in the Planck era being tiny ("reduced almost to a point") turns out to be completely erroneous. It would be truer to say that our idea of magnitude – big and small – which has developed out of our spatial experience, ceases to have any meaning in the vicinity of the Planck era. If the concepts of space and time have any meaning at all in the Planck era, they must certainly be very different from what we are accustomed to.

Along with the development of cosmology, our efforts to understand the universe have been proceeding along the reductionist and the holistic paths, until these two directions meet in the Planck era. But how did the mechanism of our understanding work? Note that the reductionist type of understanding, too, works by means of elucidating the relations between respective parts. These relations may well be oriented to "the inner bound," but if we are proceeding in this direction, we are doing so only because we are being directed along that path by our reasoning, and reasoning which is correct always follows a path of logical inferences, in other words along the relations between the premises of reasoning. The relations of logical inferences determine a certain logical structure. Therefore the understanding in question is a structuralist understanding.

Needless to say, in the physical sciences (and cosmology is one of them) the role of a network of logical inferences is performed by mathematical structures. Experimental results on their own do not allow us to understand a phenomenon; they only tell us that this is what things are like (within the limits of experimental error). Or at least would be like if there was such a thing as "experimental results on their own." We have to bear in mind that the designation of a phenomenon for observation, the design of an experiment and the apparatus needed to conduct it, the control required for it to be carried out, the discussion of errors and interpretation of results obtained are in themselves a far-reaching advance into and entanglement in a structuralist network of theoretical inferences. Nonetheless the aim of the experimental side of science is to determine the actual status quo (with all of the conditions attending it, as we have already said), while understanding – also the understanding of the experiment and its results – comes from the mathematical structure of the model or theory. The experimentally observed phenomenon becomes intelligible only when it is "fed into and read" by the appropriate mathematical structure.

Our understanding becomes the fuller the more structural relationships we discern between the various parts of a structure. Whenever we follow such a course, the given phenomenon or process reveals its dependence on other, sometimes conceptually distant, phenomena or processes, rather than just "being what it is because it is such." This holds true both for the reductionist as well as for the holistic path. In both cases the mathematical and experimental method yields understanding.

But a structural explanation cannot transcend the structure. Chains of logical inferences, be they infinitely long, will always remain within their structure. For they are the things that make up the structure. That is why a structural explanation is unavoidably committed to, and constrained within itself. If we ask for an explanation of the structure, all we get is the structure itself.

We have seen how this crucial constraint on the method works in cosmology. Can there be a better explanation for the existence of the universe than that the universe needs no explanation, since it has always existed? However, a closer look at this problem in cosmology immediately reveals a series of assumptions which have to be adopted if a model of a universe that has always been in existence is to be constructed. Strictly speaking, from the purely methodological point of view it does not matter whether we are to construct a model of an eternal universe or of a universe which had a beginning, we still have to assume some mathematical structures (mathematical formalism) to model these universes, and the following questions: where do these structures come from? Why these particular, and not other structures? And how is the transition to be made from the mathematical formalism to the real existence? In both cases they are the same questions.

And if we adopt the mathematical structures of the general theory of relativity, which lie at the basis of contemporary cosmology, then, as we have seen, the idea of an eternal universe breaks down when confronted with the theory (the problem of the cosmological constant, the expanding models) and observation (the red shifts in the galactic spectra). The universe was in a state of expansion, starting from the singularity; and new investments had to be made in order to get rid of the singularity. A variety of these were suggested: a cyclical universe, a universe with closed time-lines, the continuous creation of matter in the steady state theory. None of these proposals brought any permanent results. Not only because the results of observation turned out to be unfavourable, but also because they got tangled up in theoretical problems. In the background of all of these attempts to understand – both in the purely speculative ones as well as those which were confirmed by observation – lurked Leibniz's haunting question: "why is there something rather than nothing?" Perhaps the boldest attempt to face up to this question came in the model of the quantum creation of the universe put forward by Hartle and Hawking. But even if we admit that the mechanism proposed by this model really does produce something out of nothing, Leibniz's question is merely relegated from the realm of research in physics to the realm of the laws of nature. Nothing can be produced without the laws of nature. But why do the laws of nature exist – rather than there being nothing, genuine nothingness, with no regularities and no rationality?

The concept of an infinite number of universes does not take the edge off these questions at all, quite the contrary – it makes them all the more urgent. Although it offers an answer to the question of the special character of our universe, it calls for a justification not just for one but for an infinite number of universes. And even what it does explain is achieved at the cost of a considerable departure from the rigours of the scientific method. For we can hardly call a concept "strictly scientific" which conjures up so many existences (universes) beyond the possibility of any experimental verification whatsoever.

In spite of all these shortcomings we should not underrate the philosophical significance of the mathematical and experimental sciences, and in particular of relativistic cosmology. All the successes science has scored have been accomplished "within the framework of the method," and they are such huge successes, and they all endorse the method. The mathematical and experimental method itself has a philosophical relevance which can hardly be overestimated. For why does the universe submit to examination only if it is examined according to this method? The entire history of science shows that this is the case. Before the invention of the mathematical and experimental method the progress made in understanding the world was negligible, or rather non-existent, since all the results of any value were in fact merely steps towards the formulation of this method, if not its foreshadowing. All of this indicates that the universe has a property (or a set of properties) thanks to which it can be investigated successfully by the mathematical and experimental method, while all other methods have proved fruitless (or little better than fruitless). Elsewhere I have called this property (or set of properties) the mathematical nature of the world, devoting a considerable amount of attention to the analysis of this feature.[1] All the efforts we have made to understand the universe, as expressed in contemporary science, and especially in cosmology, have been made on the assumption that the world is mathematical. Or, in other words, all these efforts are being accomplished within the framework of the universe's mathematical structure. But the drive to understand does not stop at the discovery of the world's mathematical nature. Since the mathematical and experimental method does not reach beyond the world's mathematical nature, the drive to understand must transcend this method.

Einstein used to say that the world's intelligibility was the greatest miracle and that we would never understand that miracle. He was right insofar as in order to realign our drive to understand with that miracle we shall have to transcend the boundary of the mathematical and experimental method. If never ceasing in our drive to understand is a crucial feature of rationality, then the limits of the mathematical and experimental method are not the limits of rationality. And that is why we have to continue on the quest.

Chapter 14

~

THE METAPHYSICS AND THEOLOGY
OF CREATION

1. THE IDEA OF CREATION IN THE OLD TESTAMENT

*T*he fact that the Bible opens with a description of the creation is to a certain extent misleading. It suggests that the main message of the Bible is cosmological, or at any rate that the plot of that message is played out on the cosmological scene. This suggestion is endorsed by our view of the world, which under the impact of the progress of science in the last three centuries, sees everything from the perspective of the fact that the Earth is a small planet revolving around an average star. In addition, *Genesis*, the grandiose title tradition has bestowed on this book,[1] seems to allude to our instinct to search for roots: what is *our* genesis? However, the *Book of Genesis* is not the first book of the Bible chronologically. It was compiled during the Babylonian exile in the sixth century B.C. from passages which were probably transcriptions of a still earlier oral tradition.[2] The chronologically earlier books of the Old Testament focus on the historical aspect of God's covenant with His Chosen People (the calling of Abraham, the delivery from slavery in Egypt, the establishment of Israel as a kingdom etc.). It seems that it was not until the profound religious crisis triggered by the sack of Jerusalem in 587 B.C. and the carrying off of its people to Babylon and hence contact with a foreign culture, inspired a more outward-looking form of theological reflection. Yahweh was not only the God of one nation, but also the creator of the entire universe. Even oppressive conquerors were merely a tool in His hands. It cannot be ruled out that the high level of advancement of Babylonian astronomy also contributed to drawing the attention of the originators of the *Genesis* story to the cosmological background to their

123

M. Heller, *Ultimate Explanations of the Universe*, DOI 10.1007/978-3-642-02103-9_14,
© Springer-Verlag Berlin Heidelberg 2009

nation's history. Nonetheless, the familiar story of the Creation was not devised as a strictly cosmological doctrine, but rather as a backup to the belief that God had always been present in the history of His People.[3]

Just about everyone from our culture is (or, until recently, used to be) familiar with the opening words of the *Book of Genesis*: "In the beginning God created the heaven and the earth." Virtually all the exegetes agree that the expression "the heaven and the earth" corresponds to what we call the "universe" today, with an allowance for all the transformations the concept associated with this word has undergone due to advances in science.

The key word in the Bible's first sentence is "created" (Hebrew *bará*). On the strength of the fact that in the Bible the act of *bará* is always attributed to God, many exegetes have been trying to read the idea of a creation out of nothing into this word. However, its immediate context does not necessitate this. Emphatically, in the following sections of the account of the Creation its author (or editor) no longer used the word "created" but instead "made" (*àsá*). For instance, God *made* the firmament, dividing "the waters which were under the firmament from the waters which were above the firmament;" and He *made* "two great lights," the sun and the moon. *Bará* does not appear again until the end of the account, as it were in conclusion, in the passage about God resting after His work of Creation.[4]

The second verse of the Creation poem says that "the earth was without form, and void; and darkness was upon the face of the deep. And the Spirit of God moved upon the face of the waters."[5] This may be understood as referring either to the state from which God derived everything that exists now, or the state immediately after the original act of creation. There is only one direct reference in the Old Testament to the concept of creation, in the much later Second Book of Machabees, written in 130–135 B.C. A mother encourages her son to suffer martyrdom, saying, "and now, my son, this boon grant me. Look round at heaven and earth and all they contain [an echo of Genesis?]; bethink thee that all this, and mankind too, God made out of nothing." (2 Mach 7, 28 after the R.C. Knox translation)]. Note that these words were put into the mouth of a simple woman, therefore the "theology of the Creation" must have been a well-known truth by that time.

The phrase "In the beginning" (*bēreshit*) is not altogether clear, either. It may be understood more familiarly as "at the beginning of the world's history," or more in the context of the biblical account as "at the beginning of the work of Creation," but we should certainly not associate this expression with our present-day investigations into the beginnings of cosmic evolution. However, theologians will point to the parallel with the opening words of the Gospel of St. John: "In the beginning was the Word..." John's "In the beginning" is the Greek *en arche*, exactly corresponding to the Hebrew *bēreshit*, but it means something different from the "beginning" in Genesis. John's "beginning" is something that was before, or beyond, "any thing made that was made" (J 1,3). It is certainly right to point out the parallel between Chap. 1 of *Genesis* and St. John's Prologue, but we should not impute John's theology to the *Book of Genesis*. The two "beginnings" are cloaked in the Mystery.

2. THE GREEK CONTENTION WITH THE ORIGIN OF THE UNIVERSE

As we have seen, the truth about God creating the world had an "established position" in the Old Testament, but Jewish religious thinkers did not follow it up with profound religious reflection. For them the truth about the Creation was not so much a cosmological truth important in itself, but rather the cosmological backdrop to the history of the Chosen People, a guarantee of the fulfilment of that history. In Early Christianity the situation was quite the opposite: from the very beginning the truth about the Creation had always been a focus of attention. Of course, for Christians the "historic truths" of the Incarnation and Redemption held a central position as well, but it cannot be ruled out that perhaps they were the factors that prompted deeper reflection on God the Creator. On the grounds of a certain contrast: God, Creator of all things, entered history to become one of His own creatures: "He came unto His own, and His own received Him not" (J 1, 11). But for people living at the crossroads of the Graeco-Roman and Judaeo-Biblical cultures the truth of the Creation was a difficult truth. On the one hand it called for a re-reading of the passages from the Old Testament in a new context; while on the other hand there was a need to square up with the contribution made by the Greek philosophical tradition to thought on the origins of the world. It was a tradition constantly under pressure from evil and chaos: even if it admitted a Creator or Organiser of the world, it excluded all that was bad or disorganised from his authority. There was a certain element (matter, or perhaps chaos?) which defied the creative power of order and rationality. Christian thinkers could not be reconciled to this idea. It was from this conflict that the Christian interpretation of Creation was to develop. But before we turn our attention to this, we shall take a synthetic overview of the Greek doctrines of the origins of the world.

What we encounter here is above all a philosophical endeavour. The Greeks made a bold attempt to contend with the mystery of the origin of the universe rationally and with no overt reference to religious beliefs. I use the word "overt," since one can never rule out an unconscious or partially conscious reliance on aspects of a religious nature, even if one repudiates such a procedure. Characteristically, the Early Christian theologians who worked on the concept of Creation contended with Greek philosophy, not Greek mythology, which Christians never considered a serious partner for dialogue. From the very beginning Greek rationalism was seeping into Christian theology.

McMullin quite rightly remarks that

> *The very first philosophers of the Greek world already resorted to types of explanation of a broadly evolutionary sort. That is, they tried to explain diversity by postulating an earlier, different stage from which the present diversity developed in an intelligible way.*[6]

What is meant here is not only a search for the *arche*, the fundamental "principles" from which everything is composed (according to the Ionian philosophers these were the elements of earth, air, water, and fire, or the *apeiron*, the undetermined unboundedness); but also the opinions of those thinkers who believed in an eternal universe. They, too, were well aware of the fact that it had not always been "in its present state." According to them, the present state had emerged either due to random collisions between atoms in eternal existence (as held by the Atomists like Democritus and Lucretius), or went through cycles of change, from chaos in fire, through order, to the next catastrophic fire (as the Stoics believed).

An interesting explanation for the development of order was put forward by Empedocles, who suggested a substitute for Democritus's mechanical atoms in the form of organic miniatures of living organs able to form random combinations. But only combinations with an advantageous system had a chance of survival. Empedocles' concept was, of course, blatantly naïve, but it entailed the germ of a creative idea – natural selection.[7]

Another position on the origin of the world was represented by Aristotle, who also believed in an eternal world, but from a conceptual perspective radically different from the one held by the Atomists or Stoics. According to Aristotle the universe was eternal because the movement which made its mechanism work was eternal. But this movement was not spontaneous: it was bestowed on the world by the Prime Mover, who remained motionless himself. The Prime Mover did not operate like a mechanical cause: in his generation of movement he acted as the attractive Good. It was at this point that the Aristotelian principle of

purposefulness appeared: all beings by their nature sought to achieve the purposes proper to them, which was always a good. The harmonious order of the
world was neither the outcome of random chance, nor even of random chance
assisted by the principle of natural selection, but the result of general purposefulness. This ruled out an "evolutionary view of the world."

Plato approached the issue of the world's origin in yet another way. He
employed the metaphor of a Demiurge Artist or Artisan at work. The dialogue
Timaios, Plato's poetic vision of the creation, was most probably deliberately
fashioned as a myth to stress the metaphorical nature of the Platonic concept.[8]
His eyes set on the eternal, perfect Ideas, which incorporated mathematical
forms, the Demiurge composed the universe out of a pre-existing chaos. Here
chaos means not just disorder, as a state of suspension between existence and
non-existence, which was more in line with the Greek manner of thinking.[9]

Should we see the features of God the Creator, as Christian thinkers were
inclined to do later, in Plato's Demiurge? If we pass over the fact that, unlike the
Christian Creator, the Demiurge had to deal with the inevitabilities inherent in
the primal material, it might be worthwhile considering the interpretation
offered by McMullin, who says that, making due allowance for the metaphorical
nature of the Platonic myth, a hypothesis may be put forward that the Demiurge
was simply an image of the element of rationality at work in the world.

> *But reason is now in some sense part of the universe, just as is matter that is
> characterized by necessity. And its operation can in some sense be discerned as
> invariably present in the processes of the sensible world.*[10]

Such an interpretation (if we admit it) would be in agreement with the Greek
concept of Logos, the rationality in the world, responsible for the world's
harmony and order. The Logos concept, which went back to the times of
Heraclitus and the oldest roots of Greek thought, had already been fairly widely
disseminated by the time Early Christianity appeared on the arena of history.

3. THE CHRISTIAN THEOLOGY OF CREATION

After a short spell of wavering, Christianity transformed fairly quickly from a
splinter group in Judaism into a universalistic religion. The Christian concept of
God adapted to this transformation. Easily – because the God of the Old

Testament, though the God of the Chosen People, was also the Creator of the universe. It was enough just to draw the conclusions from this. The truth about the Creation attracted the attention of Christians not only on account of their interest in cosmology (that came later), but for its relevance to the new religion's central message. The belief that the universe had been created by an omnipotent God to Whom all things without exception were subject was a powerful demonstration of the drama in the truth of the Incarnation. Theological reflection on these truths reached its zenith in the Prologue to St. John's Gospel. The phrase "In the beginning" may be a deliberate reference to the opening words of the Book of Genesis. But John makes no mention of the Earth being "without form and void" or "an empty waste," "[i]n the beginning." In the beginning was Logos – the Word. "All things were made by Him; and without Him was not any thing made that was made." (J 1, 1-3).[11] An educated Christian of those times, brought up in Greek culture and familiar with the Bible, immediately saw a multitude of nuances in this text, which we today have to comb out with the help of meticulous comparative analysis.

Then comes a contrast: Logos the Word, the Creator of all things and present in all things, as it were became concentrated in one man. We are presented with a striking literary shortcut: "The Word was made flesh" (J 1, 14). "He came unto His own, and His own received Him not" (J 1, 11).

From the very outset Christian thought built its fundamental truths into a cosmological scene. Christ's Second Coming at the end of time was to be the finale both of that scene and the drama. The cosmological scene assumed an importance not so much on its own behalf, but only insofar as it portrayed the drama more fully.

However, the nascent Christian theology also had its practical purposes. The new dogmas had to be defended, the opposition offered by the "Greek wisdom" overcome, and answers supplied to allegations from "secular thinkers." This process was played out not only in the written and oral polemics, but also in the heads of those Christian thinkers who were true believers while at the same time immersed in the Greek culture. And this was where the original source of the conflict lay. Numerous vestiges of this have survived in the writings, as we shall see. In all the Greek concepts of the origins of the world the element of order had to contend with the element of chaos, the element of good had to overcome the element of evil (with difficulty and in defiance). Meanwhile in the Christian vision God is the Lord of all things, absolutely everything. He is responsible for everything that is and happens in the world. Not surprisingly, the Christian theology of the Creation developed in the course of a struggle to eliminate the Greek dualism of order and chaos, good and evil. That this was no easy venture is borne out by the fact that the Fathers of the Church, such as for example Justin Martyr, Irenaeus or

Clement of Alexandria, still retained a Platonic understanding of the Creation as God constructing the world out of a pre-existing matter in a state of disorder, in other words out of a matter which in a certain sense resisted the Creator.

The Early Christian thinkers would not have been men of their own times if they had not thought "in the Greek manner." That is why they could never have considered the Biblical metaphors of God's omnipotence sufficient, and sooner or later were bound to ask questions concerning "the metaphysics of Creation." The first surviving reflection on this issue is in Hermas' *The Shepherd*, an apocalyptic text written in 140–150 A.D. probably by a Greek who was once a slave. Its central subject is the problem of evil: can those who have repudiated their faith be saved? In it we read, "First of all, believe that God is One, even He who created all things and set them in order, and brought all things from non-existence into being."[12] We get a distinct stress put on "created and set in order," and – to make doubly sure – "brought all things from non-existence into being." With time the Latin phrase *ex nihilo* (out of nothing) would become a technical term.

Christianity owes its theology of the Creation chiefly to two thinkers: Origen and St. Augustine of Hippo. Origen prepared the ground, and Augustine supplied the theological finish to the concept of Creation.

4. ORIGEN

Origen (ca. 185–254) delineated a vision which is striking for its wide scope. The problem of evil and the constraints of matter play a relevant role in it as well. Originally the rational creatures had a purely spiritual nature, but they were "cast down" into the material world in consequence of their fall and turning away from God. And it was this "casting down" (*katabole*) that should be identified with the creation of the material world. But it was not merely a punishment meted out to the fallen spirits, but also a chance for them to lift themselves up and be reinstated.[13] Origen asked the following question: "What was God doing before He created the world in which we live?" And his answer was: "He was creating other worlds."

> *For Origen God the Pantocrator must always be creating, for if the world did not exist He would not have one of His fundamental attributes. Neither would there be a dimension for Him to show His love and omnipotence. Hence the (not necessarily material) world, or rather the order of creation, is a necessary product. Just as there were worlds in existence before this world, so there will be more worlds after it.*[14]

Origen stressed that God did not start being the Creator at a specific moment, and was not the Creator before.[15]

But his concept was not the Stoic idea of a cosmos in which exactly the same events are reproduced in new cycles as have already occurred previously. He made his meaning plain:

> *And now I do not understand by what proofs they can maintain their position, who assert that worlds sometimes come into existence which are not dissimilar to each other, but in all respects equal. For if there is said to be a world similar in all respects (to the present), then... everything which has been done in this life will be said to be repeated,— a state of things which I think cannot be established by any reasoning... therefore it seems to me impossible for a world to be restored for the second time, with the same order and with the same amount of births, and deaths, and actions; but that a diversity of worlds may exist with changes of no unimportant kind, so that the state of another world may be for some unmistakeable reasons better (than this), and for others worse, and for others again intermediate.*[16]

Origen's vision was too forthright and brought too many speculative elements into the Christian tradition, and hence after many years it met with firm opposition. In 553 the Second Council of Constantinople condemned several theses attributed to Origen, such as the concept of apokatastasis (viz. the return of creation to its initial state), the pre-existence of souls, and the claim that God created out of necessity.

5. AUGUSTINE

Many of the themes initiated by Origen were followed up by Augustine, who developed them in his own way. It was Augustine's edition of the theology of Creation that became the canon for later Christian thinkers. Augustine of Hippo (354–430) differed from his predecessor in intellectual temperament at least in two respects: first, he had less of an aptitude for symbolism and metaphor, and secondly, Augustine thought in Latin and not in Greek. Of course it was not just a matter of language, but above all of the cultural differences between Latin and Greek. Moreover, the 100 years exactly that separated Augustine's birth from Origen's death was a period when the Christian tradition became established well enough for Augustine to secure a firmer foothold within it. There was also another important circumstance which shaped certain relevant features of his ideas and turned out to be especially significant in his treatment of the Creation issue. Augustine had gone through a period with the Manicheans, a sect well-known for their rigorous dualism, to which even God had to be subject. This is precisely why Augustine returned time and again to questions of good and evil, freedom and grace, and the problem of Creation. Four times he embarked on extensive commentaries to the opening chapters of the Book of Genesis.

Augustine's concept of Creation was a derivative of his concept of God. Augustine's God was almighty, unchanging, and existed beyond time. The attribution of any kind of constraints to Him and removal of anything whatsoever from His omnipotence was an outcome of failure to understand His transcendence. A distinction had to be made in God's creative activity between the initiation of existence and its maintenance. "What came later" had to be understood from our point of view. There was only one act of creation in God, and it encompassed the whole of our past, present, and future. Creation was "the giving of existence" and extended over the entire period of the existence of what was created. If at any time God suspended His "giving of existence" (viz. His work of Creation) the world would immediately disintegrate into nothingness. In this sense God created the world out of nothing (*ex nihilo*).

In Book Eleven of his *Confessions* Augustine addressed Origen's question what God was doing before He created Heaven and Earth, and immediately made the following reservation:

> *I do not answer, as a certain one is reported to have done facetiously (shrugging off the force of the question). "He was preparing hell," he said, "for those who pry too deep." It is one thing to see the answer; it is another to laugh at the questioner – and for myself I do not answer these things thus.*[17]

Augustine's answer comes somewhat further down in the following words addressed to God:

> *There was no time, therefore, when thou hadst not made anything, because thou hadst made time itself. And there are no times that are coeternal with thee, because thou dost abide forever; but if times should abide, they would not be times. For what is time?*

And here Augustine gives his famous response: "If no one asks me, I know what it is. If I wish to explain it to him who asks me, I do not know."[18]

In his polemic with the Manicheans, who ridiculed the *Book of Genesis* for its inconsistent account of the Creation, Augustine worked out a principle for the interpretation of Biblical texts whenever a contradiction arose between their literal reading and "the well-established rational truth." In such cases he recommended a metaphorical interpretation. As McMullin has appositely observed, the methodological principle formulated in this way assumes that, first of all, Christian doctrine should be treated seriously as a cognitive attitude relevant to the world, and secondly that this doctrine has not been given us once and for all, but is susceptible to continuous development.[19]

How seriously Augustine treated the text of the Bible (as a source of knowledge of the world) may be seen in the following dilemma which he had to face up to. On the one hand the Old Latin translation, which Augustine used, of the Book of Sirach (Ecclesiasticus) said explicitly that "He who lives eternally created everything simultaneously" (*Qui vivit in aeternum creavit omnia simul* – Sir 18, 1). On the other hand, *Genesis* clearly suggested that the various beings came into existence gradually. Augustine resolved this problem by invoking the Stoic doctrine of *logoi spermatikoi* (Latin *rationes seminales* – seminal principles). In the original act of Creation everything was created simultaneously, but as it were in seminal form which only gradually developed later, once conditions were favourable. Augustine did not mean "seeds" in the biological sense, but philosophical principles potentially determining all future states.[20]

This doctrine, though not entirely new, since similar claims had been put forward by the Cappadocian fathers (Basil the Great and Gregory of Nazianzus), would later prove highly significant. Augustine would be subsequently cited as the precursor of the theory of evolution. This is certainly a wild exaggeration, but there is no doubt that Biblical exegesis owes some very important achievements to him.

The story is all the more intriguing as Augustine's milestone was based on an erroneous Latin translation. The Ronald Knox version of the passage reads "Naught that is, but God made it; He, the source of all right, the king that reigns for ever unconquerable." There is no mention of a simultaneous creation of all things. If Augustine had had an accurate translation available, theology would have been the poorer for one concept.

Chapter 15

CREATION AND THE PERPETUITY

OF THE UNIVERSE

1. CRISIS

*T*he formulation which left a tremendous imprint on the philosophical understanding of the Christian thinkers' idea of creation, and subsequently on the entire theological concept of creation, was the doctrine of St. Thomas Aquinas. This was of course an outcome of the enormous authority he enjoyed, and later of the fact that for a long time the Thomist philosophy was treated as well-nigh the Church's official philosophy. As regards the creation issue, St. Thomas was not an original thinker. Nearly all the components of his doctrine *de creatione* had already appeared in the thought of the Fathers of the Church and the theologians. What he accomplished was significant because he developed and systematised what had been achieved before him, and because he adapted the traditional doctrine to the needs of the times. This kind of updating of philosophical (and theological) ideas is extremely important for their continuation.

The situation St. Thomas encountered was absolutely dramatic, and it was precisely the problem of creation that was lodged in the very centre of the controversy. Prior to the thirteenth century theology had been practised in the Augustinian tradition, with strong Neo-Platonic highlights. It had elaborated its own picture of a world made up of components of Greek cosmology, and items derived from the Bible, all heavily seasoned with religious reflection which played the part of a bonding agent giving the whole the semblance of synthesis. Quite understandably, the idea of creation played a central role in that "synthesis."

M. Heller, *Ultimate Explanations of the Universe*, DOI 10.1007/978-3-642-02103-9_15,
© Springer-Verlag Berlin Heidelberg 2009

Frequently the created world would be treated as a symbol of God, which led to situations in which approaches with a mystical tendency assumed the form of philosophical or theological discourse.[1] This was prompted by the metaphorical content of Plato's *Timaios*, which constituted the principal source of information on Nature. The only commentary to the *Timaios* known at the time (and in any case incompletely), by Chalcidius, was an additional factor corroborating this trend. The Platonic myth of a Demiurge creating the world out of ever-existing, chaotic matter could be readily refashioned into the Christian version of Creation, while Plato's concept of the ever-existing ideas could be treated simply as an anticipation of the concept of the primacy of spirit over the material world.

Not surprisingly, in the thirteenth century, when Europe started to recuperate Greek and Byzantine learning based on the Aristotelian corpus of knowledge thanks to the mediation of the Arabs and with important enhancements from them, a strong reaction was inevitable. The new, more rational picture of the world posed a threat to the old image, and to the theology attached to it – or so it seemed. But the new teaching proved too much of an attraction for the philosophers and recently founded university centres for the effective defence in the long run of the "old order."

The works of Aristotle, recovered first thanks to Arabic translations from the Greek, later from the Arabic into Latin, and finally directly from the Greek, became the mainstay of the "new science," while the Arabic commentaries to them engendered a natural wave of interest. The Arabian philosopher Averroes was soon recognised as one of the best commentators. However, his assertions included not only the hypothesis of an ever-existing world, which was a reiteration from Aristotle but put more emphatically, but also other statements in conflict with Christian doctrine, such as, for example, the claim that there existed a single, collective intellect, which seemed to stand in opposition to the concept of free will. When the "Latin Averroist" movement started gaining more and more ground, its adherents appreciating the authority of Aristotle to such an extent that they became liable to allegations of subscribing to a "theory of double truths" (the truth of science and the truth of religion), a series of condemnations of the new philosophy erupted, the most renowned of which was the condemnation in 1277 by Etienne Tempier, Bishop of Paris, of 219 theses considered Averroist. One of them was the thesis of the world's existence forever.[2]

Christian thought found itself facing a crisis. It was already threatened with a disaster, which ensued four centuries later in a divergence of paths for ecclesiastical thinking and scientific thinking. St. Thomas was one of those who managed to avert the danger.

2. A PROBLEMATIC SITUATION

In Greek Antiquity the opinion that the universe had always been in existence was well-nigh instinctive. People's conviction that things had always been as they were at that moment was reinforced by the awareness that astronomical observations, which had been conducted for a long time, had not managed to discover any changes in the regular motions of the celestial bodies. That was the argument to which Aristotle resorted. The idea of the eternal existence of the universe was additionally supported by the belief held by the Greek philosophers that the heavenly bodies were composed of an unchanging and indestructible "fifth substance" or quintessence, in contrast to the four elements (earth, water, air, and fire) making up the "sublunary world." Plato had presented this idea in the *Timaios*. It was one of those concepts based on very superficial observations which stick in people's imaginations, to such an extent that they are later used to interpret and "explain" many other phenomena.

The ancient Greeks were not familiar with the concept of creation in the sense applied later by Christian thinkers. The closest to the latter was Plato's concept, according to which the Demiurge had "created" the world out of the always existing, chaotic matter. The term "created" is justified here insofar as for Plato chaos meant not so much disorder, as something on the border of existence and non-existence.

St. Thomas took a serious approach to these (and other) arguments for an ever-existing world, although he was sceptical about some of them. For example, to Aristotle's argument that generations of astronomers had failed to observe changes in the movement of the celestial bodies, he said that much more time might be required for such observations, just as no changes may be observed in a man's appearance over a period of two or three years, but that length of time was quite sufficient to observe a change in the appearance of a dog.[3]

We must also bear in mind that for St. Thomas and his contemporaries the question of whether the world had always existed or had a beginning was not just a cosmological issue, but was also integrally connected with the philosophy of God. Or, to put it more precisely, the problem of God was part of the contemporary cosmology. Hence the frequent recurrence in Aquinas' reflections of the questions whether the world was an ever-existing emanation of the Divine, and whether God could have been idle before the Creation. The former alluded to Neo-Platonic attitudes, whereas the latter was a reformulation of St. Augustine's question: "What was God doing before He created the world?" The solution to the first question was to get an appropriate definition of the concept of Creation;

the solution to the latter one was the right concept of time and eternity. This boiled down to a slightly more Aristotelian reformulation of St. Augustine's opinion: time was a measurement of motion, and since there had been no physical motion prior to the Creation, therefore time did not exist either; hence it was absurd to ask for the existence of anything at all *before* the Creation.

3. CONTRA MURMURANTES...

Among St. Thomas' many monumental works there is a short treatise which has not even been reliably dated and amounts to less that twenty pages of print, but which would be quite sufficient to dub its author an outstanding thinker. Tradition has supplemented its official title, *De Aeternitate Mundi* (On the Eternity of the World) with a sub-heading *Contra Murmurantes* (Against the Mutterers), which definitely shows that many of its readers did not take too kindly to it. Even today the main thesis proposed in this little treatise could prove a revelation for many engaged in discussions for or against Creation – if they knew of it. The thesis is as follows: we should distinguish between the idea of the creation of the world from that of its beginning; it is possible to claim without being self-contradictory that the world was (or more precisely – is being) created by God, but that it never had a beginning (in other words has always existed).

This idea was first put forward by the Jewish philosopher Moses Maimonides (1138–1204). St. Thomas' treatise might have been a reply to the sharp criticism of the Aristotelian thesis that the world had always existed, levied by St. Bonaventure and other Franciscans. This would argue for a later date (around 1270) for the treatise, although some specialists are inclined to ascribe it to an earlier period.

To appreciate the precision of St. Thomas's exposition, we first have to get through the barrage of obstacles that separates our way of thinking from the way people thought in those times. We have to realise that what we regard as thinking "within a particular system" (of Aristotelian and Christian ideas) for Thomas was simply "objective thinking." But as we get to grips with *De Aeternitate Mundi* we do not have to accept Thomas's systemic principles (although we do have to understand them) in order to see the play of ideas in his train of reasoning. Nothing helps in understanding an idea (and our aim is to understand the idea of creation) more than unravelling the chains of deduction enveloping that idea.

St. Thomas's point of departure in *De Aeternitate Mundi* is the conviction that the world was created by God. Thomas knows this as a Christian theologian, but also as a philosopher, since according to his views, the world is not a self-subsisting being, therefore it must have been given its existence, in other words created (here we are already entering the system of Thomist ideas). Thomas does not give grounds for this thesis: he did that elsewhere.[4] He has a different purpose: he wants to show that it is possible for the universe to have been created but still to have always been in existence. This can be done by showing that a "created universe" and one which "has always been in existence" are not self-contradictory.[5] In other words the meaning of these concepts has to be examined and checked to see if there is anything in them which is self-contradictory. Since Thomas and his contemporary readers had no serious problems with accepting the idea of a universe that had always been in existence, the analysis should start with a review of the concept of creation.

St. Thomas had inherited his concept of creation from the Fathers of the Church, particularly from St. Augustine, to whom he made frequent reference, and the earlier Scholastic tradition. His own contribution consisted chiefly in the expression of this concept in terms of the Aristotelian metaphysics. In that metaphysics the most profound core of existence is the substance, as is well-known. What determines the nature of a being is its substance. All the rest are merely accidents, existing only because they are "rooted" in the substance. To express the radical nature of the act of creation, Thomas says it is "the production of the whole substance of a thing."[6] God's causal creative act touches the very substance of things; without His action substance would be nothing. The consequence of this is that in the act of creation there is no "time interval" between the working of the cause and the effecting of its result, as may sometimes happen with mechanical causes. The act of creation is immediate; the creative cause need not precede its effect in the temporal sense. That is why we may envisage a created universe which nonetheless has been in existence, without risking self-contradiction. Creation need not assume a temporal beginning of the universe.

The operation of any finite cause (viz. all causes except for God) results in its effect by bringing about a change in the already existing material. The creative act does not bring about such a change; instead it brings about the coming into existence of the entire substance of a being which would otherwise not be there at all. In this sense creation is "the producing of something out of nothing." Not as if nothingness were the material from which something has been made, but because in the act of creation there is no material at all.

St. Thomas tells us that if we speak of a created world which may have always been in existence, we do not mean this in the sense of it having its existence "of itself" (thanks to itself), but in the sense that the world would be nothing if it were not for the act of creation.[7]

Towards the end of his treatise St. Thomas supports his argument with references from two authorities: St. Augustine and Boethius. From Augustine's

De Civitate Dei Thomas borrows a corroborating example: Let us imagine a man standing barefoot on sand forever. The footprint he makes in the sand is there forever, too, "but no-one would cast doubt on the fact that the footprint is being impressed by the man standing there."[8]

His reference to Boethius is to the latter's "definition" of God's eternity as "the entire and perfect possession of endless life at a single instant." A world which had always been in existence would merely be ever-existing, but not eternal, since eternity is proper only to God. His extra-temporal existence does not extend to all the passing moments of time, as would be the case with a world that had been in existence for all time. Therefore there is no danger of ascribing the divine attributes to an ever-existing world.

Chapter 16

CONTROVERSIES OVER THE
OMNIPOTENCE OF GOD

1. TWO-WAY QUESTIONS

*T*he concept of creation has a religious origin. Hence it is not surprising that it was developed and elaborated within the sphere of theological and metaphysical reflection. At any rate that intellectual environment was characteristic of the whole of the Middle Ages. But with the approach of the Modern period and the emergence of questions which would eventually give rise to the empirical sciences, the creation issue could not remain insensitive to these transformations. Although the idea of creation had become thoroughly theological and metaphysical, it obviously pertained to the world as well, and at a certain stage even served as a sort of bridgehead connecting the issues in theology and metaphysics with the gradually maturing issues in the natural sciences. Movement across this bridgehead went in both directions. Some of the theological debates left their imprint on the natural sciences, and conversely – the new style of thinking and the new methods developed by the nascent sciences generated questions addressed to the concept of creation, questions which would have been unimaginable earlier. An example of the impact of theology on science comes in a set of issues connected with the problem of divine omnipotence. What can God, and what can't He do? Is He limited by any kind of "nature of things?" Can He create something that would be self-contradictory? Or putting it more generally: is He constrained by the principles of logic? Which logic? And so on. Any constraints on divine omnipotence will of course have an impact on the created world. If there is anything that God cannot do, then that

M. Heller, *Ultimate Explanations of the Universe*, DOI 10.1007/978-3-642-02103-9_16,
© Springer-Verlag Berlin Heidelberg 2009

thing cannot occur in the created world. Can we then draw any conclusions concerning the world on the basis of limitations to divine omnipotence?

But the path of reasoning could also be taken in the opposite direction. Rapid progress in the new sciences started when they learned to apply mathematics to the examination of the world. This prompted the idea that the world had a "mathematical plan." So did the Creator think mathematically? Since from the composition of a work we may draw conclusions regarding its author's intention, we are in the midst of questions leading directly to the concept of God.

2. DILEMMAS OF DIVINE OMNIPOTENCE

In their concept of God the Greek and Roman philosophers often attained the very peak of philosophical reflection. As Amos Funkenstein writes,[1] the clash between the Judaeo-Christian and the Pagan theology did not concern the number of gods (serious thinkers treated the popular, folk brand of polytheism at best as a metaphor), but rather the nature of divinity. The Greek concept of the Divine had something about it reminiscent of Einstein's cosmic religion. For the Greeks God was a sort of cosmic principle responsible for the unchanging order of the world. The notion that God could intervene in the history of mankind or change the order of the world was unacceptable to a sophisticated Greek thinker. It is worthwhile citing the example of Galen, who ridiculed Moses for thinking that God could do anything, "even should He wish to make a bull or a horse out of ashes" if He wanted to. Galen himself believed that certain things were impossible by nature and that God did not even attempt such things at all but that He chose the best out of the possibilities of becoming.[2]

The theology of the Fathers of the Church was obliged to react to such an attitude. And it did – significantly, by endorsing it in part. Origen made a distinction between what God could do in principle (*per potentiam*), and what He actually did on a rational basis (*ex iustitia*). With time this distinction assumed the form of the classical differentiation between absolute divine omnipotence (*potentia absoluta*) and *potentia ordinata* – "ordered" power. The distinction is well illustrated in the debate between Peter Damian and Anselm of Canterbury. Peter had criticised Aristotle for the opinion that God could not change the already accomplished past *post factum*. According to Peter Damian He could, for instance make Rome never founded. Anselm immediately spotted the danger of a terrible paradox in this. If God could create self-contradictory things, He could also annihilate Himself and His omnipotence. Therefore we should assume that divine omnipotence is limited at least by the principle of non-contradiction. On the other hand, God observed the order He Himself had created, since this was His will as determined in His wisdom.

St. Thomas Aquinas made an even sharper distinction between absolute and ordered omnipotence. Absolute omnipotence pertained to everything that was not self-contradictory, and Thomas meant "self-contradiction" in the sense of formal logic; by applying a logical interpretation he stressed that any thing that accomplished such a self-contradiction would not be a thing but a non-thing, and therefore could never be created. Ordered omnipotence applied not only to the order of our world, but also to the order of other possible worlds. God chose to create this world, and not some other of the possible worlds; that was His free choice. In this sense the world was contingent – it could have been different from what it is.[3]

Of course definitions do not resolve all the problems. Duns Scotus observed that the distinction between what was possible absolutely and possible by ordination was not sharp enough, since ascribing something to God that was disordered (and that is what the distinction implied) seemed inadmissible. When we get down to specific applications the distinction becomes even fuzzier. For example, St. Thomas was of the opinion that particular beings in all possible worlds (admissible thanks to ordered omnipotence) were connected with each other by a variety of relations, and that a drastic change in any one being in such a world could lead to a logical contradiction, and therefore would be beyond the range of divine omnipotence. William Ockham disagreed with this. In his opinion every individual, as regards both its existence and its nature, was completely dependent on the will of God. The contingency of Duns Scotus' world was much more radical than that of St. Thomas's world.

The last-mentioned debate is characteristic as a testimonial to a growing voluntaristic attitude, that is a gradual expansion of the area left to the free decision of God, which emerged and increased the nearer the debaters were to the Modern period.

The peak of this trend came with the views propounded by Descartes, who said that even the axioms of mathematics depended on the will of God. If God wanted to, He could annul all the multiplication tables. This is an astounding claim, since Descartes was a resolute rationalist who believed that the whole of physics could be derived from "first principles," and considered the analytical geometry he had devised not only as a paragon of rationality, but also as a kind of ontology of the world. If the essential property of matter was its extension, as he held, then the most fundamental science of all that was material had to be geometry. Historians of philosophy and science have been racking their brains trying to figure out how to reconcile Descartes' rationalism with his radical view on the omnipotence of God.[4]

3. FROM CLASSIFICATION TO MATHEMATICALITY

In the seventeenth century the old debates on divine omnipotence found an entirely new field of application. Today we tend to think that the designers of the new science, with Galileo and Newton in their vanguard, repudiated the past and launched an entirely new style of thinking. They did indeed initiate a New Learning, but nobody is capable of abolishing his past (except in extreme cases of amnesia). Even these greatest names in science were firmly rooted in tradition, and what made them great was the fact that they did not reiterate the truths as taught to them, but were able to extract new, exciting meanings from them. Yes, they did take a rationalist outlook on the world, but they envisaged God as the guarantee of the world's rationality.

In the preceding period, going right back to Aristotle, attempts to learn about the world were made by the "categorial classification of beings." The classification of the sciences was to reflect the fundamental ontological categories,[5] and the aim of the individual sciences was to "resolve" the principal classification or to make it more detailed within the area of study proper for that particular discipline. These classifications were so natural that any change within them disturbed the world order. In this conceptual context discussions on what God could and could not do on the grounds of His ordered omnipotence seemed fully warranted.

In the seventeenth century the "categorial classifications" were displaced by the laws of nature. The question what categories of beings the world was composed of was replaced by the question how the diverse kinds of "beings" (more and more often the term "bodies" was being used) acted on, or reacted with each other. The static world was gradually turning into a dynamic world. Since the emphasis shifted to bodies in action, interest came to focus on their other properties – those which condition the action. Cassirer wrote that the old concept of substance was replaced by the concept of function.[6] To put it more pictorially, God ceased to work through the natures of things, which had been the basis of the old classification of beings; He started to work through the laws of nature. The laws of nature were constraints on the possible ways Nature may act. Nature could not do just anything at all; it had to observe its laws. The old controversies concerning constraints on divine omnipotence transformed into the observation of what Nature could and could not do. The extremities in the old debates had been voluntarism on the one hand, which made everything depend on the will of God, even the mathematical axioms, according to Descartes; and rationalism on the other hand, according to which omnipotence was restricted by aprioristic rules. In the modern version voluntarism led to an empirical approach: if God had not been restricted by any "inevitable" laws when He created the world but was free to exercise His own will, then the only way to discover the world was to open one's eyes and observe it; in other words – the only possibility for cognition of the world was empirical knowledge. A well-known hypothesis proposed by the historian of science Reijer Hooykaas says that the ascendancy of voluntarism in theology in the advent of the Early Modern period was a necessary condition for the emergence of the natural sciences.[7] But the laws of nature are expressed in the language of mathematics. That is the legacy of theological rationalism. In his opinion of the radical dependence of the mathematical truths on the will of God Descartes was isolated to such an extent that later Leibniz could write ironically:

I cannot even imagine that M. Descartes can have been quite seriously of this opinion, although he had adherents who found this easy to believe, and would in all simplicity follow him where he only made pretence to go. It was apparently one of his tricks, one of his philosophic feints: he prepared for himself some loophole, as when for instance he discovered a trick for denying the movement of the earth, while he was a Copernican in the strictest sense.[8]

The tension between the world's contingency, associated with the strategy of collecting information on the world by observation and experimentation, and the inevitability in the fact that the laws of nature are mathematical is clearly visible in the opinions of Kepler. On the one hand there was his Pythagorean belief in the world's geometrical perfection, in which the sphere and circle expressed the supreme level of symmetry; and on the other – his assiduous observation of the positions of Mars, which led to the conclusion that the orbit of the Red Planet was not a circle but an ellipse. After much intellectual struggle Kepler found a solution: things mathematical were causes of things physical, since at the beginning of time God adopted a simple but abstract plan of things mathematical to serve as the prototypes of materially designed magnitudes.[9]

Kepler's suggestion to salvage the forfeited circular symmetry by means of the symmetry of the five Platonic solids inscribed in and described on appropriately chosen circles soon lost its currency, but within a short time the idea that God was a "mathematical designer" spread and became the prevalent notion. What was still needed was to find out what that design was like. The essential core of the answer to this question, in force for the next few centuries, was supplied by Isaac Newton.

Chapter 17

～

NEWTON'S WORLD

1. NEWTON'S SCHOLIUM

*I*saac Newton earned his place in history as the creator of modern physics, but he was also a theologian. The number of works he wrote on theology is comparable with the number of his scientific works. He was a deeply committed theologian. This is documented both by his numerous statements on this subject as well as his entire oeuvre, if taken as a whole – all his achievements in physics together with their metaphysical framework. The opinions of a thinker of Newton's class must be a harmonious synthesis – at least in his own evaluation – of his diverse experiences, even if drawn from very different spheres of achievement and activity. But Newton's scientific instinct told him to keep his strictly scientific works free of his theological beliefs. Only the trained eye of the historian of science is capable of detecting vestiges of theological inspiration in works of this kind. But if that historian reads Newton's scientific works in the full context of his philosophical worldview, he will readily discern the components of a synthesis. The foundation of that synthesis was the conviction espoused by the creator of classical mechanics that the world accessible to science was not all there was to the universe. To put it in today's language, that the rationality proper to scientific method was not identical with rationality in its entirety. Moreover, looking from the perspective of the "higher rationality," in the world accessible to science one could discern vestiges of components which were inaccessible to science.

A trenchant testimonial of these opinions is lodged in the history of the *Scholium Generale*, which Newton appended to the second edition of his *Principia*. The first edition, which had no exposition on the role of God in the "system of the world" met with severe criticism chiefly from Berkeley and Leibniz. Berkeley's

145

M. Heller, *Ultimate Explanations of the Universe*, DOI 10.1007/978-3-642-02103-9_17,
© Springer-Verlag Berlin Heidelberg 2009

objection was that Newton's absolute space was either a divinity, or something infinite and coeval with God, and both of these options were absurd. Leibniz criticised Newton's concept of general gravitation as a sort of "hidden quality" which God had called into existence with no apparent sufficient reason. Roger Cotes, the editor of the second edition of the *Principia*, wrote to Newton suggesting he might answer these criticisms. That's when Newton decided to write the *Scholium Generale*, which was to be devoted entirely to the role of God in His "mechanistic philosophy." As if to justify the presence of such an extensive appendix on God in the *Principia Mathematica*, Newton added the following sentence (but only when the manuscript was going to the printers): "And thus much concerning God; to discourse of whom from the appearance of things does certainly belong to Natural Philosophy."[1]

2. A MATHEMATICAL PLAN OF CREATION

Understandably, in Newton's natural philosophy the place where the connection between his scientific "system of the world" and his vision of God is most apparent is his idea of the creation of the world. Beyond all doubt Newton was far more dependent in his opinions on the tradition out of which he had grown than we generally imagine nowadays. For instance, the roots of his concept of absolute space go back to the Scholastic discussions on the omnipresence of God,[2] while his definitions of absolute space and absolute time are a fairly faithful echo of a statement made by his teacher Isaac Barrow. As we may observe in numerous examples from the history of science, nothing crystallises a scientist's philosophical opinions as strongly as his own scientific achievements. In Newton's case, too, we see a subtle mechanism of feedback at work between tradition and his philosophical interpretation of what he had accomplished in science.

Newton inherited his concept of absolute space from tradition, but it is enough to consider only the structure of his *Principia* to realise how important a role this concept plays in his natural philosophy. It comes as no surprise that the idea passed from his natural philosophy into his philosophy of God. In the *Scholium Generale* Newton wrote the following about God: "his duration reaches from Eternity to Eternity; his presence from Infinity to Infinity." But He "is not Duration or Space," even though He "constitutes Duration and Space."[3]

In this context it is worthwhile citing a passage which states the absolute nature of space and the absolute nature of simultaneity: "every particle of Space is *always*, and every indivisible moment of Duration is *every where*," (Newton's emphasis).

Newton refutes the Scholastic concept of the omnipresence of God, whereby God is present in the world by virtue (*per virtutem*, viz. through His power), not like other bodies, which occupy a certain position in space. According to Newton, God is present in the world not only by virtue, but also in substance (viz. in His essence): "in him are all things contained and moved," even though "God suffers nothing from the motion of bodies;" and " bodies find no resistance from the omnipresence of God." Likewise, Newton departed from the Augustinian concept of eternity as the existence of God beyond time. God's existence stretched from "minus infinity" to "plus infinity," or using Newton's own words, "from Infinity to Infinity."

Absolute space and absolute time were the indispensable stage on which the world's physics was performed, but as they were the *sensoria* (organs) of the divine omnipresence and the divine eternity, they had something of the necessity of God in them. Therefore they existed, even if no processes were occurring in them. Newton formulated this clearly enough in his famous "definitions" at the beginning of the *Principia*.[4] Therefore "empty" time and "empty" space existed prior to the beginning of the world. When He created the world, at a particular point in absolute time God called into being bodies endowed with mass, placed them in particular points in absolute space, and gave them their initial velocities. Ever since that moment the laws of motion Newton discovered had been taking the world in the direction determined by them. In other words, the only difference between Newton's idea of the creation of the world and the solution of Cauchy's problem for the equations of motion (viz. the setting of initial conditions for these equations) was that the former applied not to a physical sub-system, but to the entire universe and that the "setting" of the initial conditions involved not only the determination of the initial positions and velocities bodies had, but also the calling of those bodies into existence.

Newton was in no two minds that the solution to Cauchy's problem for the whole universe required a Rational Cause. In reply to Canon Bentley's questions, he wrote in a letter:

> *To your second query, I answer, that the motions which the planets now have could not spring from any natural cause alone, but were impressed by an intelligent Agent.*

Somewhat further on he added:

> *To make this system, therefore, with all its motions, required a cause which understood and compared together the quantities of matter in the several bodies*

of the sun and planets, and the gravitating powers resulting from thence; the several distances of the primary planets from the sun and of the secondary ones [moons] from Saturn, Jupiter, and the earth; and the velocities with which these planets could revolve about those quantities of matter in the central bodies, and to compare and adjust all these things together, in so great a variety of bodies, argues that cause to be, not blind and fortuitous, but very well skilled in mechanics and geometry.[5]

According to Newton God the Creator behaved like a mathematician and His work followed mathematical principles.

3. PHYSICO-THEOLOGY AND THE CONCEPT OF CREATION

It is widely held that as soon as He created the universe and entrusted it to the laws of mechanics, the Newtonian God ceased to be interested in it. The laws of mechanics were sufficiently precise to govern the world on their own. This is a later idea; Newton himself was far from such an opinion. Indeed, for him the laws of mechanics were sufficient to secure a defined course for the world, but they were entirely dependent on the will of God. Not only could He have created any laws of nature He liked, but He could also intervene in their operation or suspend them completely according to His will. In the *Scholium Generale* Newton stressed that God was the Lord, Ruler, and Pantocrator of the universe. God ruled the universe not as one rules one's own body (pantheistic tendencies were appearing already), but as a Sovereign Prince. But God was not an abstract principle, entirely external with respect to the universe. He was a personal and rational being; yet it should be remembered that His eternity and omnipresence constituted absolute time and absolute space – the arena in which the laws of nature operated. In this sense the universe existed in God.

Newton belongs to the voluntaristic tradition in English theology. Moreover, he reinforced that tradition very considerably. Thanks to his enormous scientific authority his numerous greater and lesser epigones would continue that tradition, especially as, apart from his theological reasoning, Newton himself invoked arguments associated with his scientific achievements. He was of the opinion that every so often God intervened in His "system of the world." This happened, for instance, when comets visiting the Solar System caused too great a perturbation in its previously synchronised motions and there was a need to administer new initial conditions.[6] Soon the search for this kind of "gap" in current scientific theories became fashionable. It came to be held that the chief task of natural theology was to explain such "gaps" by the direct intervention of God. Today this trend is referred to as physico-theology, and is considered to have been based on both a theological error and a methodological error. Leibniz pointed out the theological error when he wrote that the Newtonian God had not been provident enough to have created a perfect work which required no amendments.[7] The methodological error soon came to light once the progress made in science gradually started to fill in the gaps in hitherto current theories, and the "God hypothesis" was no longer necessary. Today physico-theology and "resorting to God to fill in gaps" is censured (by theologians as well), but we must remember that at the time it was perhaps a more or less inevitable outcome of the first phase of progress in science, and of the over-enthusiastic reaction to the avalanche of successes accumulating at such a rate that it was hard to imagine that the avalanche did not conceal many still unexplained areas.

As regards the opinions of Newton himself the situation was even more subtle and connected with his idea of creation. Note that the gaps in our knowledge need not be located in the middle of the history of the universe (e.g. corrections to the movement of the planets); they may lie at the beginning, where they are harder to spot. The Newtonian creation concept itself has the nature of a gap in our knowledge: for the equations of motion require the "setting" of initial conditions in order to govern the world. But since Newton's theory is not capable of setting the required initial conditions, the task is assigned to the Creator.[8] At least Newton was consistent: once he decided to introduce the "God hypothesis" to his cosmological vision, on subsequent occasions (in corrections), he was less hesitant to do so, while his voluntaristic theology made him more confident.

4. NEWTON'S IMPACT

It would be hard to overrate Newton's impact on later developments in science. This fact is so obvious that we may consider it generally known. But he also exerted an enormous influence on the development of theology, both in the textbooks as well as of the popular kind. Today few people are aware of this fact. The tremendous prestige of classical physics meant that soon a vision of the world based on it became the prevalent view, and certain aspects of Newton's personal opinions were inseparably associated with it. Post-Tridentine catechisms adopted this view, tacitly and perhaps not fully aware of doing so, in their formulas. For anyone relying on such catechisms for their religious instruction it was well-nigh obvious that whenever we spoke of the creation of the world, what we meant was that at a certain time in His eternal and unending existence God called the world into being and ruled it according to His will. The course of the world in accordance with the laws of nature was the natural sphere; while the actions of God transcending those laws were the supernatural sphere. The average believer envisages even God's hearing of his prayers as a divine "adjustment" to what was to occur.

Such views became so ubiquitous that even those who, in increasing numbers, rejected and contested them on the grounds of atheism, argued against the Post-Newtonian version of popular theology rather than against the interpretations proposed by the Grand Masters of traditional theology. The anti-religious slogans of the French Enlightenment and thereafter of Positivism went hand in hand with a continuous lowering of standards in theology.

Chapter 18

~

LEIBNIZ'S WORLD

1. NEWTON AND LEIBNIZ

S ummarising Isaac Newton's philosophical views is a relatively simple matter. Although they are scattered throughout his greater and lesser works, letters and treatises, they were crystallised fairly quickly and were always consistent later. Newton was precise and well-ordered not only in his works on physics and mathematics, and his philosophical commentaries were associated with his physics in a fairly natural manner: as soon as one manages to comprehend his interpretative principles it is not difficult to reconstruct the ideas espoused by the pioneer of classical physics. With Leibniz the situation is completely different. Certainly he had no want of genius, but he was occupied with too many matters, not only ones connected with science and philosophy, to concentrate resolutely and systematically on any one subject. He devised calculus as if incidentally; he had so many excellent ideas in physics but never formulated them systematically; he created his metaphysical system while busy with numerous other activities and engaged in several debates. His only major work, *Theodicée*, is more of a collection of essays than a systematic treatise. Leibniz's philosophy is original, profound, and it staked out the paths for future developments, but it is not easy to interpret. The first difficulty is that his texts may be selected and arranged in diverse ways. Their chronology is not always the best guide to them. The history of the continuators of the ideas initiated by the Great Librarian of Hanover shows that those ideas may be read in a variety of ways. This certainly does not mean that Leibniz is a "dark" philosopher and may be understood in any arbitrary way. His ideas have clearly defined boundaries (beyond which there is only territory alien to him), easily distinguishable from the views of other philosophers, but it is precisely because of the abundance of his

151

M. Heller, *Ultimate Explanations of the Universe*, DOI 10.1007/978-3-642-02103-9_18,
© Springer-Verlag Berlin Heidelberg 2009

ideas that they may be pursued and taken in different directions, and in point of fact one never knows which of these directions was the one Leibniz himself chose – or would have chosen.

All of this induces me to refrain in this chapter from giving a faithful account of Leibniz's views. After all, I'm not writing a textbook on the history of philosophy; my aim is to follow the development of the concept of the ultimate comprehension of the universe and its Christian version, that is the concept of creation. The ideas held by other people are only a guide to my own deliberations. In this chapter, more than in the preceding ones, I shall allow myself to present my own interpretation. Of course I shall try to "keep to Leibniz," but following only those things in his works which I have discovered for myself.

2. WHEN GOD CALCULATES AND THINKS THINGS THROUGH

If I had to choose one statement out of all the writings of Leibniz which gives the fullest expression of his idea of the creation of the world, my choice would be the following sentence which Leibniz wrote in the margin of a text entitled *Dialogus*[1]: "When God calculates and thinks things through, the world is made."[2] But, as is usually the case with formulations that say it in a nutshell, you have to put a lot of effort and profound attention into unravelling the full sense of this sentence. The rest of this chapter will essentially be a commentary on this sentence by Leibniz.

Every one of us has had some experience of calculation. Whenever the numbers are not too big calculation is mechanical, done almost without thinking, and once you master the basic techniques of calculation you can also say the same thing of operations carried out on big numbers. Real mathematical thinking only starts when you have to solve a more complicated problem, or formulate and prove a theorem – in other words, whenever you have to find a mathematical structure, understand the way it works (for mathematical structures are not static, even if they do not change with time!), construct a new structure starting from a given one, and see its relationship with other structures. . . Manipulation of this kind with structures is usually associated with calculation, or lead to calculation, since mathematical structures tend to dress up in numbers and the language of calculation is their natural language.

This is the sort of image we should attach to Leibniz's expression that when God "calculates and thinks things through" the world comes into being. The Latin verb phrase *cogitationem exercet*, translated as "thinks things through," literally means "exerts his thinking" or "applies his mind." To understand what Leibniz meant, we may imagine the work he must have done to devise calculus – differentiation and integration. First of all he had to identify the problem, assemble the components which eventually went into his solution but were scattered throughout the work of his predecessors, make a few crucial approximations, prove a number of theorems which express the relationships between the components of the new structure, perform the calculations for many examples, formulate new computational procedures, and finally show that the structure which emerges as a result of all these steps works perfectly when applied to the theories of physics.

Now Leibniz's metaphor of God creating the world by calculation and the exertion of His mind is more comprehensible. All we have to do now is to liberate it from all the human limitations and imperfections, and supply it with an important corollary: that for God to obtain a result means that the result has come into being. This intuition, too, will be more readily comprehensible if we imagine a mathematician at work on a new scientific theory. Whenever a mathematical structure is devised for application in physics, its definitions are selected in such a way as to correspond to the experimental results, the structure's most readily comprehensible components are given an appropriate interpretation, sometimes with a certain amount of modification to make them work better. And when thanks to all these steps the structure has reached the required level of maturity – hey presto! The mathematical structure turns into a theory of physics. Not only does it explain something we have already observed in the world, but it also predicts new, sometimes very subtle phenomena.

But when God calculates and exerts His mind, there is no trial and error, no fitting together of this or that. The universe simply comes into existence.

3. SECRETS OF THE DIVINE CALCULATION

With the help of Leibniz, let's now try to look into God's way of thinking. At the beginning of the *Theodicée* Leibniz writes that "reason is the linking together of truths" and he immediately adds that "this definition of reason (that is to say of strict and true reason) has surprised some persons."[3] Let's admit that it has surprised us, too, but we should realise that what Leibniz had in mind was reason as "the content of the mind." Several pages later we read

> *For I observed at the beginning that by REASON here I do not mean the opinions and discourses of men, nor even the habit they have formed of judging things according to the usual course of Nature, but rather the inviolable linking together of truths.*[4]

The word "link" deserves to be emphasised. What Leibniz had in mind were not "truths" in themselves, but also the connections between them by chains of deduction. At any rate, according to him, such is "strict and true reason." Elsewhere Leibniz stresses that "Right reason is a linking together of truths, corrupt reason is mixed with prejudices and passions."[5]

How does human reason differ from God's reason? Leibniz makes the following remarks:

> *Reason, since it consists in the linking together of truths, is entitled to connect also those wherewith experience has furnished it, in order thence to draw mixed conclusions; but reason pure and simple, as distinct from experience, only has to do with truths independent of the senses.*[6]

God simply encompasses all the possibilities (Leibniz speaks of "possible worlds") and all the logical relations between them. These relations have the nature of mathematical deduction. This is what Leibniz's "When God calculates..." means.

Human reason has something of this "divine spark" about it; it is able to make its way along the deductive chains linking the various possibilities, albeit to a limited extent. In his *Monadologie* Leibniz writes:

> *... what distinguishes us from the lower animals is our knowledge of necessary and eternal truths and ... gives us reason and science, raising us to the knowledge of ourselves and of God.*[7]

Thus, by analysing the way we reason we may attempt to understand the way in which God "calculates and thinks things through," the way He creates the world.

According to Leibniz there are two main principles which govern the way we reason: the principle of contradiction, "on the strength of which we judge to be false anything that involves contradiction, and as true whatever is opposed, or contradictory to what is false,"[8] and the principle of sufficient reason,

> *on the strength of which we hold that no fact can ever be true, or existent, no statement correct, unless there is a sufficient reason why things are as they are and no otherwise – even if in most cases we can't know what the reason is.*[9]

There are two kinds of truth corresponding to these two principles: truths of reasoning, true on the grounds of the principle of contradiction: they are "necessary, and their opposite is impossible;" and truths of fact, which are true on the grounds of the principle of sufficient reason. If something cannot be determined on the principle of contradiction, we must search for a sufficient cause why it is so and no otherwise. Since the world is the result of God's calculation and planning it is as rational as possible and contains nothing for which there is no rational cause.

There may be situations in which there are an infinite number of deductive steps from the premises to the conclusion. But that is no obstacle for God to recognise such a conclusion as a necessary rational truth. Of course human reason is not able to cover the infinite distance separating the premise from the conclusion; for human reason such a conclusion may be only a truth of fact. In such a situation human reason can search for a sufficient reason why things should be so and no otherwise. As Leibniz writes, "God has given us a concession," allowing us to give a sufficient reason to justify what we are unable to deduce on the principles of logic.[10] When confronted with a contingent truth our procedure could consist in decomposing reasons into their component elements, but in this way we shall never obtain a full proof. In such cases only God, "Who alone in a single spiritual glance comprehends the infinite chain of causes," understands the "reason of the truth."[11]

Does this mean that in His recognition of necessary truths on the grounds of deduction God Himself is subject to necessity? Yes, but He Himself is that necessity: "...necessary truths depend solely on God's understanding, of which they are the internal object."[12]

Leibniz defines the world (or universe) as everything that exists (except for God). Hence by definition there is one universe. Yes, there exist other worlds, but only potentially, in the mind of God.

> I call 'World' the whole succession and the whole agglomeration of all existent things, lest it be said that several worlds could have existed in different times and different places. For they must needs be reckoned all together as one world or, if you will, as one Universe.[13]

Of the infinite number of all possible worlds God has chosen one, the one which exists in reality. What determined His choice? Since He is the Best and Most Rational Being, He chose the best of all possible worlds.

Ever since the times of Voltaire to the present day and many of the contemporary thinkers, this point in Leibniz's doctrine has been notoriously ridiculed. If this is the best of all possible worlds, they say, then what would the worse ones be like? However, this line of criticism is facile, not based on a profound appreciation of what Leibniz meant. God's selection of the best possible world is like the process of optimisation in mathematics.

> As in mathematics, when there is no maximum nor minimum, in short nothing distinguished, everything is done equally, or when that is not possible nothing at all is done: so it may be said likewise in respect of perfect wisdom, which is no less orderly than mathematics, that if there were not the best (optimum) among all possible worlds, God would not have produced any.[14]

The key link in Leibniz's reasoning – and one that is not always well understood, it seems – is that he was not thinking of a best possible world in the absolute sense, that is best for all of its components regardless of every other component. There is simply no such thing as a best possible world in this sense, since the very concept is self-contradictory. And according to Leibniz a self-contradiction can have no corresponding reality, it is just nothing. The world is a system of components associated with each other by a variety of relations, and we may only speak of a best possible world in the sense of the best possible system of the entire network of relations. To put it in a simplified way, we may say that it is a question of the good of the entirety with the smallest possible infringement of the good of its individual components.

> For it must be known that all things are connected in each one of the possible worlds: the universe, whatever it may be, is all of one piece, like an ocean: the least movement extends its effect there to any distance whatsoever, even though this effect become less perceptible in proportion to the distance.[15]

Every perturbation, even a small one, in every possible world, causes a consequence even in the remote parts of the whole. The evaluation of the optimum must take into account all the possible perturbations of this kind.

It might be worthwhile here to quote Max Planck, another thinker evidently fascinated by Leibniz. In an article on the principle of least action Planck wrote

In this context one may certainly recall Leibniz's Theodicée, *which contains the theorem that of all the worlds which could have been created the real world is the one which, apart from the unavoidable evil, contains the most good. This theorem is nothing else but the variational principle, in a form which is exactly the same as the form of the subsequent principle of least action. The inevitable intermingling of good and evil plays the role of imposed conditions, and it is obvious that all the properties of the real world, down to the details, could be derived from this principle if a rigorously mathematical formula could be found to express the measure of good on the one hand, and on the other the imposed conditions. The latter are as important as the former.[16]*

Let's take a look at this crucial point in Leibniz's doctrine from a somewhat different perspective. I have said that according to Leibniz whatever is self-contradictory is equivalent to nothingness: whatever is self-contradictory cannot exist. In this sense God "must contain fully as much reality as is possible."[17] And conversely: "a thing's perfection is simply the total amount of positive reality it contains."[18] In this sense

all creatures derive from God and from nothingness. Their self-being is of God, their non-being is of nothing. . ... No creature can be without non-being, otherwise it would be God.[19]

If in His calculations God has selected the best possible world, then it is the world which contains the least amount of nothingness and the maximum of existence.

Leibniz would probably have said that those who make fun of his arguments exist because nonetheless they must be contributing in some way or other to the maximisation of the good of the entirety.

4. TIME AND SPACE

Leibniz's concept of creation as the calculation and thought of God carries obvious consequences for his other ideas of the world, particularly of time and space. They found their fullest expression in his polemic with Clarke, who represented the views of Newton, so essentially this was a debate between Leibniz and Newton. Leibniz could not accept the Newtonian concept of absolute time and absolute space, capable of existence without any events, prior to the creation of the world. They would have been perfectly homogeneous, and therefore would not have marked out any point in space or time. Therefore God would not have had any reason to choose a particular point in time and space rather than any other to create the universe. We shall demonstrate this argument by referring to a passage on time:

> ... *Suppose someone asks, "Why didn't God create everything a year sooner than He did?" sees that this has no answer and infers that God has made a choice where there couldn't possibly be a reason for His choosing that way rather than some other. I say that his inference would be right if time was some thing distinct from things existing in time, or events occurring in time, for in that case it would indeed be impossible for there to be any reason why events shouldn't have occurred in exactly the order they did but at some different time.*[20]

Somewhat earlier there is an analogous passage on space.

Time and space are not "external with respect to things;" according to Leibniz they are relations arranging things or events in order (a thing may be regarded as a particularly enduring collection of events). Leibniz wrote:

> *For my part, I have said several times that I hold space to be something merely relative, as time is; taking space to be an order of coexistences, as time is an order of successions.*[21]

Relations arranging events in an order of successions one after another constitute time; while relations arranging events such that they are "coexistent" constitute space.[22]

If time and space are the relations between events, then they cannot exist if there are no events. Therefore God did not create the universe *in* time and space, but *with* time and space. In this sense Leibniz returned to St. Augustine's concept, but with a new, fuller validation. The world is not so much a collection of objects, but rather – to use the language of today – a structure, in other words the set of relations from which objects derive their essentiality.

5. CAUSALITY

If the universe is the result of divine calculation, then it is the work of the Mathematician and is mathematical itself. Today the mathematical nature of the world is understood to mean the assertion that there is an astonishing correspondence between the structure of the world and certain mathematical structures: a correspondence so astonishing that more information on the world may be obtained more efficiently by the study of a given mathematical structure than by the laborious collection of experimental data. In any case, in advanced theories of physics it is impossible to design any experiment without resorting to a highly developed mathematical apparatus. Of course experimentation is a salient part of a research strategy, if only to obtain confirmation that we have selected the right mathematical structure for the examination of the given part of the world.[23] Although in Leibniz's days mathematical (theoretical) physics was only at the beginning of its spectacular career, his genius showed an amazing grasp of this extraordinary method. Let us, for instance, scrutinise the following sentence: "the Region of the Eternal Verities must be substituted for matter when we are concerned with seeking out the source of things."[24] "Eternal Verities" is a term used by St. Augustine, and for Leibniz it means "mathematical beings" (or in a more modern expression "mathematical structures"). According to the notions prevalent at the time, physics was to study the material world, but here was Leibniz saying that anyone who wanted to study the world at its source should "substitute" mathematical structures for matter.

The modern version of the concept of the world as a mathematical structure is usually associated with mathematical Platonism (though there is no inevitable link between these two doctrines), in other words with the belief that mathematical structures (or mathematical objects) exist objectively, independently of the human mind and aprioristically with respect to the physical world. In this sense Leibniz was undoubtedly a Platonist, but a special type of Platonist, who held that mathematical beings exist in God and draw their power from Him. We could say that he was a Platonist of the Augustinian type.

Mathematical Platonism is a doctrine which is fairly widespread today among mathematicians and physicists engaged in philosophy, but not so popular with philosophers of physics. An objection which the latter group often raise against mathematical Platonism is that mathematical beings cannot exert a causal impact on the world, since they are beyond the world. For example, Michael Dummett writes that abstract objects, which is what mathematical beings are, do not have "causal power" and therefore "cannot explain anything, and . . . the world would appear just the same to us if they did not exist."[25]

Leibniz not only anticipated this criticism, but also reversed it: the world ("matter") does not explain anything of itself; to find the "source of things" we must turn to mathematical structures. Matter has no "causal power;" all causality comes from mathematics, to which matter is "subordinated." To refer to a modern example, when a particle of cosmic rays collides with atoms in the upper atmosphere and produces a cascade of other particles, the reason why this happens is not because a mathematical structure gives an approximately accurate description of this process, but because the particles are the implementation of a given mathematical structure and do exactly what is encoded in that structure. Leibniz would have said that if there were no mathematical structures there would be nothing.

Chapter 19

THE INITIAL SINGULARITY AND THE
CREATION OF THE WORLD

1. THE QUESTION OF EVOLUTION AND ITS BEGINNING

*I*n the previous chapters we carried out an overview – albeit briefly – of various ideas of creation. We saw that it is a wide-ranging concept, and comes into play in many different opinions and controversies in philosophy, theology, and even the natural sciences. It is certainly not a static concept and has been making an active contribution to the development of doctrines and visions of the world. The question arises what the contemporary advances in science and the critical review of those advances have added to the career of the concept of creation. Only now are we putting the question directly, but both the question itself as well as an attempt to answer it have been present in this book well-nigh from its first page. The doctrine of the creation of the world seems to call for confrontation with at least two features of the world image as delineated by contemporary science. These two features are the evolutionary nature of the picture of the world, and the problem of the origin of cosmic evolution, better known as the question of the initial singularity. Let's start with some general remarks on evolution, then (in this chapter) ask about its origin, and later (in the next chapter) return to a consideration of the relation between evolution and creation.

M. Heller, *Ultimate Explanations of the Universe*, DOI 10.1007/978-3-642-02103-9_19,
© Springer-Verlag Berlin Heidelberg 2009

2. TIME AND ITS BEGINNING

The system subject to evolution is a system which changes with time. The conceptual problems start already at this point. As we recall, the arena of what happens in the universe is not time and space taken separately, but the conjunction of time and space in one structure called space-time, which, as we known, is static, viz. all of it exists all at once. Of course we may impose conditions on space-time to make it resolved into a single time and distinct spaces, one for every moment in time, but the possibility of such a resolution is a very special exception, not the rule. Moreover, for the abstract space-time to become the space-time of a specific universe it must be a solution of Einstein's equations and satisfy strictly defined conditions (see in Chap. 4 Sects. 4 and 5). The overwhelming majority of solutions do not permit a resolution of space-time into a single time and momentary spaces, and those that do constitute a "zero measure" subset in the set of all possible solutions. A perusal of the textbooks on the theory of relativity gives the false impression that quite the contrary is true. This impression is due to the fact that on the whole physicists are not interested in solutions which do not have a single time and do not study them. This is because the universe in which we live, or at least that part of it which we can survey by observation has a single universal time. One of the great successes of the standard cosmological model is that it has managed to reconstruct the history of the universe from the first moments of the Big Bang up to the present day. That history is measured out by a single, universal time. Again it turns out that we are living in a very exceptional universe. The adherents of the anthropic principle gain yet another argument. In a universe without a single time it would be hard to imagine an evolution long enough to have led – from the primordial quark soup through the nucleosynthesis of the chemical elements and the chemistry of carbon, the development of galaxies, stars, and planets – to the emergence of biological evolution and the origin of life.

The geometry of the standard cosmological model predicts the existence not only of a universal time, but also of its beginning. Time emerges in the form of the initial singularity, which we came across on many occasions in the previous chapters. For instance, in Chap. 3 we saw how the existence of the initial singularity cast doubt on the viability of a cyclical cosmological model, and in Chap. 7 we had the opportunity to see what kind of "quantum tricks" Hartle and Hawking had to resort to in order to get rid of the initial singularity. We know from our deliberations so far on the various concepts of creation that the creation of the world does not necessarily mean the same as its temporal beginning, but since many authors have simply ignored this principle, it might be worthwhile to look at the initial singularity in relation to the idea of the world being created by God.

3. PROBLEMS WITH THE SINGULARITY

The initial singularity may be regarded as the mathematical equivalent of the Big Bang. When the first relativistic models of the universe were constructed in the 1920s and were compared with observations soon afterwards, it turned out that the universe was expanding and the galaxies moving away from each other at continually accelerating velocities. It became obvious intuitively that the universe's phase of expansion must have started from something like a gigantic explosion. Originally this was called "the initial fireball;" only later was the "Big Bang" adopted as its name. In the first cosmological models this super-dense beginning was described by means of a theoretical expression: "the volume of the universe tends to zero and its density tends to infinity as time tends to '$t=0$'." Later the term "initial singularity" was established for this "limiting process." It became natural to identify the physical intuitions about the Big Bang with the mathematical intuitions associated with the initial singularity, and it all linked up in a suggestive picture of a beginning of the world.

Intuitions and suggestive pictures were enough to set off the debates on world view which soon erupted. Some wanted to see the initial singularity as the moment when the world was created, while others tried to disavow such a conclusion by constructing a variety of rival models, such as the cyclical universe (Chap. 3), closed-history universes (Chap. 4), or the steady-state cosmology (Chap. 5). But intuitions and suggestive images are not enough for rigorous research. As soon as the question was put how to remove the initial singularity from the cosmological model it turned out that first a proper definition had to be given for the singularity. An expression like "as time tends to '$t=0$'" is not a good definition, as in most cosmological models there is no universal time. The problem turned out to be quite formidable, and it was not until the early 1960s that not so much a definition of the singularity was put forward, but rather a criterion whereby it was possible to establish whether there existed a singularity for the given space-time.[1] In the theory of relativity the histories of particles and photons are represented by curves in space-time – the histories of particles by time-like curves, and the histories of photons by zero curves (we met time-like curves in Chap. 4, in the discussion of Gödel's cosmological model). If the histories of all the particles and photons, in other words all the time-like and zero curves, in a particular space-time may be extended indefinitely in both directions,[2] then in that space-time there is no singularity.[3] If it is impossible to extend the history of at least one particle or observer indefinitely, then there is a singularity in that space-time; its history "breaks down" at the singularity. Note that it is not a question only of the initial singularity; the history may break down at the final singularity into which the entire closed universe collapses, or at a singularity in the centre of a black hole.

The formulation of this criterion soon led to the proof of some important theorems concerning the existence of singularities. The proof was arrived at by Penrose and Hawking (as well as others), as we mentioned in Chap. 3 Sect.6. There is a whole collection of these theorems and they all have a similar structure: if certain conditions are met in a given space-time, then in that space-time there exists a singularity (viz. at least one time-like or zero curve breaks down in it). The various theorems in the collection formulate a variety of conditions which must be satisfied for a singularity to occur, but in general these conditions are "natural," that is they are fulfilled in space-times that are "physically realistic."[4] The general conclusion from these theorems is that the occurrence of singularities in cosmological models is the rule rather than an exception, and that they cannot be eliminated by simple means from those models in which they do occur.

At the turn of the 1960s and 1970s, when the theorems concerning singularities and the conclusions resulting from them were in the vanguard of research on the general theory of relativity, there was not much interest in quantum cosmology. It was generally assumed that the theorems relating to the existence of singularities resolved the controversy in favour of those who argued for a beginning of the world. Stephen Hawking and George Ellis concluded their fundamental and highly technical monograph on the singularity problem with a celebrated declaration:

The creation of the Universe out of nothing has been argued, indecisively, from early times... The results we have obtained support the idea that the universe began a finite time ago. However the actual point of creation, the singularity, is outside the scope of presently known laws of physics.[5]

The last sentence is not just a reservation made by physicists who do not want to stray into areas which are beyond their bound. It has its grounds in the formalism of the theory. If the criterion for the occurrence of a singularity is the breaking down of the history of particles or observers, then we know nothing about the nature of the singularity other than that it is a borderline at which our knowledge comes to an end.

However, soon after the publication of Hawking and Ellis's monograph relativists started to become more and more interested in the quantisation of gravitation and the consequences of this operation for cosmology, in other words in quantum cosmology. Hawking himself addressed the problem (one of the outcomes was the quantum model of the creation of the universe he and Hartle constructed: see Chap. 7). From the outset it was known that the singularity theorems were applicable only to "classical singularities," viz. when the quantum effects of gravitation were not taken into account. As soon as interest was aroused in the search for a quantum theory of gravitation, the philosophical significance of the singularity theorems ceased to be so obvious. The quantum effects of gravitation may break one of the conditions in the singularity theorems, hence there might be no need for the occurrence of a singularity in the history of the universe.

Whether that will be so or not will depend on the future theory of quantum gravity. As yet we have not developed such a generally accepted theory. The various theories and models which have been proposed offer different answers to the question. Though most of these models do away with the singularities. But this might be a "selection effect," since in general physicists tend to look for models with no singularities. However, it cannot be ruled out that the future theory of quantum gravitation will bring a tremendous surprise, with conclusions nobody is expecting today. A thorough grasp of the history of science teaches us to be prepared for such surprises.[6]

4. METHODOLOGICAL RESERVATIONS

From what we have said so far it is clear that the question of the initial singularity is neutral with respect to the creation problem, for scientific, methodological, as well as theological reasons. Let's review the scientific reasons first.

If we disregard the quantum nature of gravitation,[7] we still do not know anything about the physical nature of the singularity, as we recall. All we know is that at the singularity there is a breakdown in all the information that we may obtain on the universe on the grounds of the non-quantum laws of physics. Metaphorically speaking, even if there was something "before" the singularity, the world is oblivious to it (I have put "before" into inverted commas because at the singularity time breaks down and the concept of "before the singularity" has no sense). Perhaps the future quantum theory of gravitation will change the situation. Moreover, we have reasons to expect it to do so, but until it does we cannot really draw any definitive conclusions.

The methodological grounds are even more cogent than the scientific ones. The fundamental rule in the methodology of science is never to stop in our research efforts, never to call it a day and say we can go no further. In Chap. 1 we called this attitude the totalitarianism of scientific method. The task of science is to "explain the universe by means of the universe itself," and not to invoke causes which lie beyond the universe. But we must bear in mind that this is a *methodological* principle, that is it should be treated as an assumption concerning the method of research, not as an ontological certainty. After all, God might as well have created the world out of nothing last night, along with all the tree-lines and fossils ready-made and testifying to their antiquity. He might have encoded in our minds a memory of events that never happened. But science may not take such "miracles" into account. Even if the future theory of quantum gravitation confirms the existence of a "strong singularity" at the beginning of the present phase of the universe's evolution, it will not pass on such a conclusion to the theologians for further processing, but will face a new scientific challenge.

Finally there are the theological reasons. The question of the initial singularity is certainly a gap in our present scientific knowledge. But filling up this gap with God would be a return to the strategy used in the age of physico-theology (see in Chap. 17 Sect. 3), subsequently labelled rather ironically the God-of-the-gaps method. Contemporary theology should not stray back into that historic error.

All the more so as there are even deeper theological reasons. The overview of a selection (not all) of the philosophical and theological concepts of creation we carried out in Chaps. 13–18 showed that the creation idea does not fully overlap with the concept of the beginning of the universe. The creation concept may entail the idea of a beginning (although it need not from the philosophical point of view), but it is far ampler. Moreover, in the course of history the human view of the universe and its creation has evolved. The extension of our cosmological horizon and progress in science have played a significant role in this process. In times past God was Lord of the World, that is the Earth, while the stars were not much more than a trimming; nowadays we imagine God in terms commensurable at least with the Cosmos as we understand it today, and when we think of creation we subconsciously assume this means the creation of all that contemporary cosmology is about. Meanwhile one of the principal theological truths is that God is a transcendental being, that is He transcends all that we are capable of imagining. Thus we should bear in mind that His act of creation entailed much more than what is accessible to our contemporary scrutiny. In Chaps. 8–12 we discussed the concept of many universes, perhaps an infinite number of them. A scientist may have serious misgivings as to whether this is a scientific concept or not, but a theologian should definitely take it into consideration. If God is infinite, then – as someone appositely observed – He may not be interested in anything that is less than infinity. Furthermore, a good theologian has grounds to think that the created reality is far richer than what we are able to ask about in our inquiries.

5. THE GREAT SIGN

Does that mean we should give up our quest for "signs of creation" in the world that surrounds us? Of course not. The only problem is that the whole universe is the work of God and if by "sign" we mean something like a footprint on a sparkling stretch of snow, in other words a local effect indicating that there has been an intervention, then in point of fact we are looking for a gap to be filled with the God hypothesis. Putting it in another way, the entire universe is one great sign of God, so no wonder we're looking in vain if our attention is concentrating on details which are to point to the existence of a Creator. So let's focus on the Entirety. One may certainly contemplate in silent awe the immensity of the galaxies, the profundity of space, the aeons of time, the abysses of the black holes and the gigantic energy of the Big Bang. That's what some of the poets and writers of science books for non-specialists do. Maybe it's worthwhile taking an occasional look at the artistry of the Grand Cosmos. But such experiences can only be an "effect of the scale": on the scale of the discoveries made by science we are but a negligible speck of dust. That's why it's quite easy to feel a respect for the Immensity. But it's enough to remind oneself that the Immensity may itself be just a speck – not much more than nothing – in the infinity of other universes, for the respect to turn into bewilderment and a feeling of hopelessness.

What we should be doing when we focus on the Entirety is not to allow ourselves to be carried away by our emotions, but rather to search for something that characterises the Entirety at its most profound level. We have been looking at something like that from several vantage-points virtually from the first page of this book. What I have in mind is the *rational aspect of the Entirety*, the property thanks to which we are able to examine it rationally. All the scientific theories, all the controversies concerning their interpretation, all the philosophers' deliberations have thrived thanks to this aspect of the Entirety. If there were no rational element in it any answer to any arbitrary question put to the world at large would be equally good, while a universe rent apart by contradictions could never have come into existence at all.

The Entirety is the Great Sign of God – the Mind of God, the Creative Concept inscribed in the existing universe. All the scientific theories, all the efforts to arrive at the right interpretation of them, all the endeavours philosophers have made constitute the collective attempt of humankind to decipher the Concept inherent in the structure of that which exists.

Slowly the components of that Concept are beginning to fit into an Entirety. There are still many cracks and crevices in it, but they are the outcome of our lack of proficiency, not of the Concept. We have the right to think that by keeping to the scientific method and refraining from premature conclusions we shall gradually fill in the crevices and seal up the cracks. Often what we discover seems so exciting that we lose the strength and will to take it one step further.

That Grand Concept has one more property. We do not know in what language it has been written, but from our point of view as its rather clumsy interpreters it appears to have been written in the language of mathematics. We ourselves have created this language, chiefly to crack the code of Nature (and thereafter as it were as an art for art's sake), and it has turned out to be so successful that we have the right to believe that in a way it reflects the Language of the Concept itself.

Perhaps it's as Leibniz suspected: when God engages in His Mathematics, there arise universes.

Chapter 20

CREATION AND EVOLUTION

1. TWO MISAPPROPRIATIONS

*W*ords are not innocent, like for example chairs or pebbles on the beach. Words can injure or kill. Words can be misappropriated and instead of saying what they mean can start misleading, leading astray. Something of this sort has happened recently to two verbal expressions: "creationism" ("creation science") and "intelligent design." The term "creationism" was well-grounded in the philosophical and theological tradition and was in standard use in the sense of the Christian doctrine of the creation of the world by God. But the word has been misappropriated by fundamentalist groups in the USA who are convinced that the admission of the biological theory of evolution is in conflict with the Christian religion, claiming that the literal interpretation of Chap. 1 of the Book of Genesis should be accorded the name "creation science" and calling for "equal rights" for this "science" with the theories of modern biology. It took quite a long time for these ideas to reach Europe from the United States, nonetheless now, when someone (even in Europe) admits to a belief that the world was created he is almost automatically branded a Fundamentalist. I am not going to argue with "creation science." Anyone who has read the previous chapters will see how profoundly unwarranted that standpoint is.

When the Fundamentalist Creationists started losing case after case in the American courts for "equal rights" for their ideas with "official science" they adopted a new strategy. They dropped the "creation science" label and the direct reference to the Biblical account and transferred to a more philosophical dimension. The theory of evolution explained the origins of life and its subsequent development by a series of random events. But, they said, in His creation of the world God implemented His "intelligent design," and random occurrence conflicted with "intelligent design."

M. Heller, *Ultimate Explanations of the Universe*, DOI 10.1007/978-3-642-02103-9_20,
© Springer-Verlag Berlin Heidelberg 2009

Rather than wrangle with the theory of evolution, they now said it should be reshuffled and augmented by emphasis on the vestiges of Intelligent Design in Nature. This doctrine was "more intelligent" than Fundamentalist Creationism, and many religious thinkers fell for it. An attentive reader who has gone through the previous chapters of this book would no doubt manage to answer this line of argument on his own, but it might be worthwhile to take a closer look at this issue. Yes, I shall embark on polemic, but not so much in order to refute the arguments of the "opposition," rather to treat these claims as an opportunity to scrutinise some of the aspects of the theology of creation.

Yet again the term "intelligent design" has been misappropriated by a group of dissidents (with respect to authentic science). When He created the universe, God applied a transcendent counterpart – meaning a counterpart transcending all our concepts – of what we rightly call "intelligent design." No theologian would quarrel with such a statement. But today, if we resorted to the expression "intelligent design" we would, more or less automatically, be banded together with the group of "scientific dissidents." That is why in the previous chapter I did not use the expression "intelligent design" but "Creative Concept" (compare with Einstein's idea of the "Mind of God"). Generalising, the semantic difference between these two terms is that whereas "intelligent design" assumes an opposition between God's design and random occurrence, "creative concept" makes no such assumption. In this approach God is Lord also of random events, which He incorporates in His Creative Concept. I am of the opinion that an idea to the contrary is a serious theological mistake.

2. THE HYPERSPACE OF LIFE

Let's start with some remarks on the theory of evolution and its place in the general system of science. For all the sciences about the world constitute a system. Of course the different sciences are concerned with different areas or aspects of the world, but they are all connected not only by sharing the same elements of scientific method, but also by the fact that they are all committed to the study of the same world, thanks to which the results obtained in one discipline of science may – and often do – carry consequences for other scientific disciplines. This applies especially to biology and physics. Living organisms are undeniably also physical bodies. But the dependence goes deeper. Living organisms could not exist without organic chemistry; organic chemistry is the chemistry of carbon compounds; and as soon as we put the question of the origin of carbon (and we can hardly not put that question), we are in the realm of astrophysics, or even cosmology.

Today we have a fully developed theory of nucleogenesis (viz. the origins of the nuclei of the chemical elements) which agrees well with observations. These observations consist in the comparison of the level of agreement between the predictions of the theory of nucleogenesis with the abundance of the particular elements in the universe now. We know, for example, that the nuclei of the light elements were synthesised within the first few minutes after the Big Bang, when the temperature in the universe was high enough to facilitate the nuclear reactions which gave rise to the emergence of atomic nuclei. The nuclei of deuterium (the heavy hydrogen isotope), helium and lithium were made in this way. All the other chemical elements were made much later in the interiors of massive stars. This also applies to carbon, the crucial element for the emergence of life. But for carbon to be made, the original hydrogen had to be "burned through" in several generations of stars. Massive stars end their lifecycles in an explosion known as a supernova, and new generations of stars arise from their ashes. This cycle has to last for some 9–10 billion years. In a time as long as this the recession of the galaxies will have expanded the universe to a distance of some 9–10 billion light years. Thus, for carbon-based life to appear on at least one planet on an orbit around a star, the universe must be old and big. In this sense life is a phenomenon of cosmic significance, even if it exists on just one planet.

Biological evolution is undoubtedly a complex dynamic process, and as such is embedded in the dynamics of the universe; it is one of the strands of that dynamics closely bound to its other strands. Let's take a closer look at the extraordinary fine tuning of the life of which we are carriers with the structure of the Entirety.

The basic building blocks of life are molecules of the amino acids. There are 20 different standard amino acids from which proteins are synthesised. Let's assume that one protein consists of a hundred such molecules (not too far-fetched an assumption). On the basis of this assumption it may be readily calculated that there are 100^{20} (i.e. 10^{39}) different combinations possible – a mind-boggling number. This is also the estimate of the total number of electrons in the universe. The well-known biologist Simon Conway Morris invites us to imagine that all of these combinations make up a gigantic "hyperspace" in which each combination marks a different point, and he asks how many of these combinations lead to the emergence of life. All the proteins extant on Earth account for only a microscopic part of that huge space. What are the chances of a random movement taking us to that sub-domain? He writes:

> *Despite the immensity of biological hyperspace I shall argue that nearly all of it must remain for ever empty, not because our chance drunken walk failed to wander into one domain rather than another but because the door could never open, the road was never there, the possibilities were from the beginning for ever unavailable.*[1]

In other words, the "drunken walk" which was to take us to the "domain of life" was not so "drunken" after all. Most of the routes to the empty domains were simply blocked. Putting it less metaphorically: yes, chance played a role in the emergence of life, but some outcomes were more probable than others. God cast the dice (another metaphor!), but His dice were loaded. The idea of "loaded dice" (weighted probabilities) fits comfortably into the laws of physics; if the probabilities had been slightly different, the universe would have remained forever barren, inhospitable to life. To put it in another way: random events are part and parcel of the laws of physics, which are written in the language of mathematics, and probability theory, which governs chance and random events, is a mathematical structure. So no wonder that along with other mathematical structures it is part of the sophisticated composition that makes up the software of the universe.

3. PROBABILITY AND CHANCE

Let's take a closer look at the mathematical structure responsible for the "standard" concept of probability (there are other, "non-standard" concepts of probability as well in mathematics). If we give some more attentive thought to the matter, we shall admit that the attribution of a probability to a variety of occurrences is reminiscent of taking a measurement.[2] For instance, when I measure the length of the table I find that its length is a certain number of units (e.g. 2.5 m). If I say that my lottery ticket has a 1/3,000,000 chance of winning, I'm also ascribing a number to a particular event. Here it doesn't matter that it's an event and not an object like a table. What's more relevant is the fact that I'm ascribing a number between zero and one to events. If I didn't buy a lottery ticket the probability of winning would be zero; if I bought up all the tickets, the probability of winning would be one (but then I would have to pay more for all the tickets than the jackpot was worth). In all other cases the probability would be a fraction between zero and one.

This example is instructive, as in mathematics probability theory is a special case of the theory of measure. We won't go into the technical details; here all we have to do is remember that mathematics is a formal science which tells us nothing about the world. Hence the mathematical measure theory is not concerned with real measurement processes; it only formulates the principles for the attribution of numbers ("measures") to certain subsets of a given space and deduces conclusions from those principles (axioms). If we impose a "normalising condition" on this measure theory, viz. a condition that the sum of all the measures (the sum of all the numbers ascribed to the subsets of the given space) must be equal to one (and all measures are positive), then we say that the measures are probability measures and that under this condition measure theory is probability theory.

Note that in probability measures understood in this way there is no sense of uncertainty, hesitation or anticipation of the kind we tend to associate with the concept of probability. There are only hard and fast rules, and their principles of operation – just as there are in every other branch of mathematics.

How does all this relate to "estimating the probabilities of various events occurring in the real world"? In just the same way as for other cases whenever we apply a mathematical theory to the observation of the world. We have to "apply" the given mathematical theory to the world, viz. *interpret* it as a structure of the world, or – in other words – acknowledge that the given mathematical theory is *a model of the world* (usually only in a certain respect). Of course we do not do this arbitrarily, but by applying standard research procedures devised by science, that is above all we try to take heed of the verdict of observations and experimental results. The results obtained hitherto, in combination with the history of the given problem, usually suggest which mathematical structure we should adopt, and later the comparison of the predictions obtained on the basis of the mathematical model we have constructed with subsequent observations and experimental results will tell us whether to accept or reject the model.

It's the same with probability theory. The mathematical measure theory itself does not tell us anything about the world until we interpret it, that is "apply" it to the world. Let's take a very simple example. Tossing a coin. Heads or tails? If the coin is true we say that the probability of heads (or tails) is ½. What does this mean? From the mathematical point of view what we have here is a space consisting of two subsets, one of which is labelled "heads" and the other "tails". To each of them we ascribe the measure ½. This is a probability measure, because the sum of the measures for all the subsets is equal to one. Now we say that this basically very simple mathematical structure is the model of the physical process of tossing a coin. But we have to verify the model empirically. We carry out a long series of tosses and make a record of the number of heads and the number of tails. If in a long series of tosses the result is approximately "one to one," that is half the results are heads and half tails, and the longer the series of tosses the better the approximation, we have the right to say that our model is working properly.

Note, however, that whether or not the model is good depends on the world and what it is like. In the mathematical probability theory we may ascribe any measures we like to the various subsets, provided their sum equals one. But it is observation, in other words what the world is like, that determines whether the selection of the measure of ½ for each of the two subsets is correct.[3]

Let's now take a look at this in the light of the concept of creation as presented in the previous chapters. God thinks mathematically. In creating the world He implements certain mathematical structures as the structure of the world (of course the mathematical structures we have discovered are merely very rough approximations to the structures which God uses in His thinking). The fact that probabilities observed in our world assume specific values is part of God's "creative concept."

Let's now go back to the question of chance as God's "competitor" or "rival." If by chance or random occurrence we mean an event with a very low probability which nonetheless does happen, or in other words an event to which we should ascribe a low probability measure in the given set of events, then in the light of what we have said above such an event is still part of God's "creative concept." Therefore chance events are also "fully controlled" by God.

Thinking of random occurrence in opposition to God is tantamount to treating standard probability theory as an absolute, that is putting it above and beyond God's control. We feel intuitively that anything that occurs frequently (viz. it is highly probable a priori) does not need to be explained. But anything that happens rarely (viz. it is not very probable a priori) is either a chance occurrence or else has been specially contrived by someone. But as we have seen, a low probability is not a sort of anti-absolute in opposition to God, but part of His creative strategy. Furthermore, there are other probability theories in mathematics different from the standard one described in this chapter. For instance, in statistical quantum mechanics and quantum field theories a generalised probability theory is used,[4] and recently free probability theory has been developing rapidly.[5] And it is by no means obvious which of them (or perhaps some other probability theory) will be applicable at the deepest level of the structure of the universe.

4. GOD AND CHANCE

In Christian theology random occurrence has never been set up in opposition to God. This finds its expression in the popular kind of devotion, for instance when someone thanks God for saving his life by making him held up in a traffic jam and miss boarding a plane which later crashed killing everyone on board. But isn't such an attitude in conflict with the idea of a Divine Plan in the work of creation? Doesn't the admission of genuine random occurrence kill the idea of planning in the bud?

We should bear in mind that our notion of planning is imbued with our general experience and our immersion in the flow of time. The pursuit of an aim implies a decision to select that aim in a future context and the initiation of a series of actions to accomplish that aim. The full prediction of the outcome of such actions will only be possible if the process is entirely deterministic and its development is not susceptible to small changes in the initial conditions. In such situations the occurrence of a chance event will undermine the possibility of accurate prediction. But this is a highly anthropomorphist understanding of planning. To show the possibility of an alternative understanding it's worthwhile referring to the Augustinian or Leibnizian concept of an extratemporal God. Ernan McMullin puts it appositely:

Terms like "plan" and "purpose" obviously shift meaning when the element of time is absent. For God to plan is for the outcome to occur. There is no interval between decision and completion. Thus the character of the process which, from our perspective, separates initiation and accomplishment is of no relevance to whether or not a plan or purpose on the part of the Creator is involved.[6]

In the last sentence of this quotation McMullin means that from God's point of view of planning a process it does not matter whether the process is deterministic or "interspersed" with random occurrences, since God does not deduce the final state from previous states but knows it "by inspection". There is no expectation in His planning. What for us is an element of chance, brutally intervening in the course of events (as predicted by us), for God constitutes an element of the "composition of the world."

Note that this understanding of the "Creation Plan" removes yet another objection frequently invoked against the theological opinion that God knows the future. It is often claimed that if God knows the outcome of my future actions, then that outcome has been determined before I accomplish it. Hence God's foreknowledge cannot be reconciled with my free will. But if God knows events which from my point of view are in the future not on the grounds of deduction but by inspection, then I can be the free agent of an action which God has always seen from His extratemporal perspective.

This philosophy of contingency, and the planning and accomplishing of a purpose has obvious consequences for the debates going on concerning "intelligent design." We shall let Ernan McMullin take the floor again:

It makes no difference, therefore, whether the appearance of Homo sapiens *is the inevitable result of a steady process of complexification stretching over billions of years, or whether on the contrary it comes about through a series of coincidences that would have made it entirely unpredictable from the (causal) human standpoint. Either way, the outcome is of God's making, and from the Biblical standpoint may appear as part of God's plan.[7]*

Thus Christian doctrine may be reconciled with a variety of interpretations of the origin of "novelties" like life or consciousness, but theology, in other words the rational interpretation of religious truths, is obliged to take into account the well-grounded results of science. And in this respect the verdict of science is clear enough: the universe in which we live is an evolutionary process, one of the strands of which leads from the primal plasma through the synthesis of the chemical elements, the emergence of the galaxies, stars and planets, to the inception of biological evolution and the flourishing of self-awareness. Any theological system which ignores that grand Cosmic Symphony condemns itself to marginalisation and chooses to follow a road leading to nowhere.

Chapter 21

⁓

LEIBNIZ'S QUESTION

1. L. KUHN'S CATALOGUE OF EXPLANATIONS

*I*n No. 2 of Volume 13 of the well-known journal *Skeptic* there is an article by Robert Lawrence Kuhn entitled "Why This Universe? Toward a Taxonomy of Possible Explanations."[1] Kuhn is patently excited by the prospects opened up by contemporary theoretical physics and cosmology. These prospects transcend the method employed in the empirical sciences in the narrow sense, well-nigh compelling the more inquisitive mind at least to ask questions. The anthropic principles have drawn attention to the exceptionality of our universe within the space of all the possibilities, and the idea of a multiverse has thrown the gate open to speculation. Kuhn decided to compile a "taxonomy" of all the explanations various authors have put forward for the amazing fact that the universe we live in is what it is and no other. As one reads Kuhn's "catalogue of explanations," which to a large extent overlaps with the explanations presented in the previous chapters of this book, one develops the impression that for Kuhn the question why the universe is what it is was a surrogate question. His real question comes at the beginning and end of his article. In his introduction Kuhn admits that already when he was twelve he was suddenly struck by the question *why there was something in existence rather than nothing*. This admission is followed by an italicised paragraph which I shall quote *in extenso*:

> *Why not Nothing? What if everything had always been Nothing? Not just emptiness, not just blankness, and not just emptiness and blankness forever, but not even the existence of emptiness, not even the meaning of blankness, and*

M. Heller, *Ultimate Explanations of the Universe*, DOI 10.1007/978-3-642-02103-9_21,
© Springer-Verlag Berlin Heidelberg 2009

no forever. Wouldn't it have been easier, simpler, more logical, to have Nothing rather than something?[2]

This question haunted Kuhn. In his catalogue of explanations it is implied rather than expressly formulated. "Why is the universe what it is?" is only the inevitable sequel of "Why is it at all?" The question has accompanied us throughout this book, and now, at the end, we cannot but put it outright.

In the first part of this book we saw that cosmology, the contemporary science of the universe, cannot break free from asking ultimate questions. Admittedly, more insistent versions of such questions transcend the borders of the mathematical and experimental method employed in cosmology, but the representatives of this science often cross these boundaries themselves and indulge in speculation that is not so constrained by methodology. Nonetheless in all of these speculations there has to be a point of departure; you have to make some initial assumptions: maybe mathematics, the rules of deduction, the laws of nature, an infinite number of universes... If your initial assumption is NOTHING, then you stay with NOTHING forever. That is why I decided to write the third part to this book, on the concept of creation. It is a concept which attempts to face up to the question of "Why something rather than nothing?" Yes, it does go beyond the mathematical and experimental method, it cannot be any otherwise, but it has an established place in the history of European philosophy. So if I have devoted a large part of this book to an attempt to answer the question "Why something rather than nothing?" why am I returning to that question again in a separate chapter? Partly to dot the i's and cross the t's, but above all to take a look at the attempts to dodge the question.

2. LEIBNIZ'S QUESTION

As I have said, the question of why something rather than nothing has been present in Christian thought from the very beginning, but this form of the question and its dramatisation comes from Leibniz. A word of explanation is needed for the expression "dramatisation." Leibniz formulated his question in a fairly short treatise entitled "Principles of Nature and Grace, Based on Reason,"[3] and he did so dryly and with no drama whatsoever. But apparently the contrast between the brevity of the question itself and the intensity of the sense – dramatic in itself – that this question carries fixed itself so firmly in the memories of the following generations of thinkers that afterwards they were never able to ask for the reason of the existence of anything at all other than in the way Leibniz had done. So let's consult the original text. After rather briefly introducing the reader to the main ideas in his monadology, Leibniz states:

> *So far I have spoken only of* what goes on in the natural world;[4] *now I must move up to the metaphysical level, by making use of a great though not very widely used principle, which says that nothing* comes about without a sufficient reason...[5]

As we remember from Chap. 18, the principle of sufficient reason (along with the principle of contradiction) is what determines the whole of Leibniz's thought. And it makes him ask the following question:

> Why is there something rather than nothing? *After all,* nothing *is simpler and easier than* something.[6]

Leibniz's answer to this question is perhaps rather hasty (at least so it seems if read without the context of his other works), and may seem to us too flimsily grounded. According to Leibniz the universe is made up of a "series of contingent things," hence

> *a sufficient reason that has no need of any further reason – a "Because" that doesn't throw up a further "Why?" – and this must lie outside the series of contingent things, and must be found in a substance which is the cause of the entire series. It must be something that exists necessarily, carrying the reason for its existence within itself.*[7]

Perhaps that is precisely the fate of that question: every unsatisfactory answer makes the question become more and more vexing.

3. THE DOMINO EFFECT

There have been several attempts to "neutralise" Leibniz's question. I shall present a few of them. Here is the first, frequently invoked in discussions.

We shall simply have to reconcile ourselves not so much to there being no answer to this question, as to there being no possibility of obtaining an answer to it. Expecting an answer to Leibniz's question would mean calling for the deducing of something from non-existent premises. You can hardly hold it against the logician if he is incapable of doing that.[8] Quite apart from the fact that all you can deduce from premises, i.e. a set of statements, is another statement (e.g. saying that something exists), but not the fact of the existence itself of something. However, if we gloss over this logical slip, the above attempt to neutralise Leibniz's question is in fact its cogent, really dramatic reformulation. On the one hand we have nothing, zero existence (and no premises, either, to deduce anything from);[9] while on the other hand there is the undeniable existence of something – the universe. We can't hold it against the logicians that they cannot tackle this problem. Putting it somewhat metaphorically, the problem is the infinite distance separating NOTHING from SOMETHING. What Leibniz was asking was how to cross that distance.

The author of the entry in the *Stanford Encyclopedia of Philosophy* I have just quoted writes with what looks like a touch of irony that there is not much consolation in Hume's observation that although we can't explain the existence of all things, we can explain the existence of every thing separately. Imagine an infinite row of dominoes standing upright and then tumbling in an avalanche-like manner, each pushing down the domino behind it. We know what caused the fall of each domino, even though we don't know what made the whole row start to tumble. Hume was being optimistic in trusting that we are able to explain the existence of each thing on its own. Science tries to explain "the existence of every thing," certainly not "on its own" or "separately," but in far-reaching association with other things. And alas, as we have seen in this book, it is still a long way off from the final success. But it has scored some remarkable successes "on the way." Which makes Leibniz's question even more dramatic: not only should we be asking why something exists, but also why that something is open to rational methods of examination. This is the source of Leibniz's principle of sufficient reason – only in a rational world can we ask for reasons, also for the reasons why anything at all exists.

4. THE EXISTENCE OF THE UNIVERSE AND THE RULES OF LANGUAGE

One of the variations of the above objection invokes the "philosophical syntax of language." Leibniz's question is a combination of words which are meaningless. The syntactic error consists in the fact that the word "nothingness" does not refer to anything and we can neither ask a sensible question about "nothing," nor can we say anything at all about nothingness. Also the question "Why does something exist?" carries a syntax error, since it assumes that there exists a "something else," apart from "something," which could explain that "something."[10]

The dispute over the philosophical (or logical) syntax of language separates off analytical philosophers from practically all other trends in philosophy, and even within the analytical fold there is no unanimity on this issue. Hence resorting to the rules of logical syntax in order to neutralise Leibniz's question has the character of an "intra-systemic" criterion; outside the system it is not regarded as legitimate. I am far from querying the achievements of analytical philosophy, in the field of the philosophy of language as well. But it is one thing to determine the principles of "philosophical grammar," and quite another to apply them to specific cases.

Philosophical issues certainly have a "linguistic component," and ignoring it is a serious fault on the part of many philosophers. Every philosopher should be analytical as regards this component. But then resolving philosophical problems, including the Great Philosophical Problems (and Leibniz's question is one of them) solely by means of linguistic resources is a serious fault on the part of many (not all) analytical philosophers. Often such solutions consist in getting rid of the problem as meaningless. One may not assume a priori that everything that cannot be formulated in ordinary language, even as rigorously defined as the language of philosophy, is not a genuine problem. Try formulating an advanced mathematical structure, e.g. describe the structure of spinor space, in ordinary though rigorously defined language. It is self-evident that mathematics is the language that has been created specifically for the description of structures like spinor structure,[11] though this in no way alters the fact that spinor structure is a good example to show the limitations of ordinary language.

It is good to bear in mind Quine's warning. After a rather arduous analysis of certain ontological problems he wrote: "But we must not jump to the conclusion that what there is depends on words."[12]

5. THE PROBABILITY OF NOTHING

Peter van Inwagen proposed a rather peculiar answer to the question why there exists anything at all.[13] His reasoning is as follows. There may exist an infinite number of worlds full of diverse beings, but only one empty world. Therefore the probability of the empty world is zero, while the probability of a (non-empty) world full of beings is one.

This apparently simple reasoning is based on very strong and essentially arbitrary assumptions. First of all, that there may exist an infinite number of worlds (that they have at least a potential existence); secondly, that probability theory as we know it may be applied to them (in other words that probability theory is in a sense aprioristic with respect to these worlds); and thirdly, that they come into being on the principle of "greater probability." The following question may be put with respect to this mental construct: "why does it exist, rather than nothing?"

In fact once we have put this question we could consider our discussion with van Inwagen finished. However, I cannot refrain from referring the reader back to Chap. 20 Sect. 3, where I argued that we should not treat probability theory as an absolute and turn it into an ontology which governs everything, even the decisions made by God. Probability theory is simply a very good mathematical theory and the fact that it may be successfully applied to the world is truly astonishing. Should there be any readers with problems in accepting this statement, I encourage them to re-read Chap. 20 Sect. 3. Let's consider the example of throwing a true die (see in Chap. 12 Sect. 3). In connection with van Inwagen's argument, let's ask what is the probability of throwing none of the numbers. The answer is self-evident: there is no such possibility at all. But why? Because we ourselves have defined the distribution function for the probabilities, on the grounds of many experiments, for the set of all possible outcomes of throwing the die. That function assigns the same probability, 1/6, to each of the possible outcomes, that is 1, 2, 3, 4, 5, and 6. On the grounds of the definition of the distribution function we have ruled out the occurrence of any other outcomes except for the above-mentioned ones. Of course we could have given a different definition of the probability function, but it would not be applicable to the throwing of a true die. For instance, it might apply to the throwing of a die with bevelled corners, in which case we would have the grounds for a definition of a probability function with a value assigned for the probability of not throwing any of the numbers.[14]

Rather that treat probability theory as an absolute, it might be worthwhile to stop and think for a moment how it actually works.

6. A BRUTE FACT

There is one further way out of the situation. To present it, I shall refer to Helena Eilstein, who in her recent book *Biblia w ręku ateisty* (The Bible in the Hands of an Atheist) made her position plain.[15] In her introduction she writes that she considers herself an atheist, not an agnostic, and gives an extensive explanation that there are many ways in which a given attitude may be rejected. She also explains in what sense she rejects the belief in the existence of God. It's an interesting question, but not so relevant to our reflections right now. What is of interest to us is the manner in which someone who denies the existence of God tackles Leibniz's question.

Helena Eilstein starts her "approach" to the question with a remark that every scientific hypothesis which is to explain something is based on "certain assumptions" which are treated as "given" and in themselves not subject to explanation. Sometimes an explanation may be obtained for them thanks to subsequent theories, but it may happen that "their explanation is beyond the cognitive powers of the human intellect."[16] There follows a cogent observation:

> *In fact, one of the characteristic features of contemporary science is that the limitation of the human cognitive powers is becoming more and more comprehensively apparent. Our observations cannot encompass the universe, irrespectively of whether it is constrained in terms of space-time or not. Our experiments cannot "directly" reach all the layers of physical existence, because, for instance, it is impossible for us physically to achieve the energy necessary for this. Moreover, sometimes it happens in science that asking for an explanation becomes warranted cognitively only once we have achieved the capacity to obtain an explanation.*

Eilstein "extrapolates" these undeniably true observations back to a more extreme case:

> *We cannot rule out that some of the givens relied on by science are unexplainable for ontological reasons; they are "ontologically primary" and therefore do not call for an explanation, but merely for confirmation.*

Note that the supposition that there are certain problems which science will never solve (and certain facts it will never explain) is quite natural, and many scientists and philosophers concur; but the claim that some of these problems relate to "ontologically primary givens" is a very strong ontological assertion.

Eilstein gradually approaches the central issue:

> *In the scientific presentation of reality we should take into consideration the inevitability of having to acknowledge the conjecture that in certain of its most*

essential aspects the universe simply is what it is, and that we shall have to base
our scientific explanations on this.

She calls answers to questions why this or that thing exists "existential explanations." An existential explanation may refer to the laws of science or the initial conditions for the given issue. The property that all explanations, including existential explanations, have in common

> *is that they take for granted that something exists and is what it is, and that this*
> *acknowledgement needs no further explanation, at least within the bounds of the*
> *given explanatory procedure.*

Again a relevant observation, but it should be supplemented with the remark that science never withdraws from the possibility of explaining what has been accepted as "initially given" in the explanations obtained hitherto. In the opinion of many, even "the ultimate theory" will not bring an end to questions.

And for this reason what Helena Eilstein goes on to write may not be inferred from these remarks. She continues in this way:

> *From the above it may be inferred that the question why something exists rather*
> *than nothing is illegitimate. The question is illegitimate since by the very nature of*
> *things there can be no answer to it. The fact that it exists is the ultimate, brute fact.*

In the original Polish text Eilstein adds a footnote to explain that she had the English expression "brute fact" in mind, but could find no good Polish equivalent (she uses the phrase "naked fact"). Sympathising with her translation problems, I would recommend following the phonetics and writing *brutalny fakt* ("brutal fact"). Indeed, for anyone concurring with Helena Eilstein's opinion, the existence of anything whatsoever is a brutal fact – brutal because it violates the principle which for me is the expression of rationality: that we should go on asking questions for as long as there is still something left to explain. And in philosophy it often happens that even if there is no answer to some questions, their examination may lead to progress.

EPILOGUE: THE LESSON OF
PSEUDO-DIONYSIUS

~

The mysterious author of a work entitled *The Divine Names* claims that God is inaccessible to human cognition, and that the Divine Names appear to us merely as His shadow. Paradoxically, he devised a name for himself which has successfully concealed his true identity. When the Apostle Paul preached in the Areopagus of Athens, the Greek men of learning listened to him attentively for as long as he spoke of the Unknown God. But as soon as he mentioned the Resurrection, "some mocked: and others said, We will hear thee again of this matter." Only a few showed an interest in Paul's teaching. One of them was Dionysius the Areopagite. So much on the subject in the *Acts of the Apostles* (17, 16–34). An unidentified author, most probably a fifth-century monk who called himself Dionysius the Areopagite wrote a few original theological works. It was not so much an appropriation or hiding behind somebody else's name, rather an act of modesty in line with the custom of the times on the part of this author, who did not want to steal the limelight but instead to endorse what he had written with the authority of someone more widely known. Today we refer to him as Pseudo-Dionysius the Areopagite.

Pseudo-Dionysius was an unusual author. On reading his texts we have mixed feelings. He is terribly "out-of-date," embroiled in Neoplatonic deliberation, mystical (in fact he coined the word "mysticism"), with an excessive predilection for classifying the Choirs of Angels. It takes a certain amount of patience and a sense of taste to discover in his works theological reflections worthy of a master. And it is to one of his themes that I would like to devote some attention at the end of this book, which has been concerned with the

185

M. Heller, *Ultimate Explanations of the Universe*, DOI 10.1007/978-3-642-02103-9,
© Springer-Verlag Berlin Heidelberg 2009

struggles that have been going on with the most difficult questions that can be put to the universe.

The word "struggles" seems particularly apt. In the first part of this book we reviewed a series of heroic attempts to "explain the universe by the universe itself." In spite of numerous spectacular successes "on the way" the undertaking ended in ... an opening up on further attempts. In the book's second part these attempts took the form of speculations on the anthropic principles and an infinite space of universes. I think that the principal message of that part of the book is that in the search for answers to ultimate questions it is hard to get away from infinity. In the third part of the book we considered the philosophical and theological idea of creation. The question of an ultimate explanation for the universe was answered, but at the cost of being immersed in the Infinity of God. And so we have come up against the Mystery, or rather – as we ourselves are part of the universe – we have let ourselves be overwhelmed by the Mystery. Not without misgivings or opposition. We have seen how some of us have resisted the Mystery by invoking lesser mysteries: the brute fact of existence or the intricate relations between the syntax of language and reality. Letting oneself be overwhelmed by the Mystery, even with misgivings – that is the problem addressed by Pseudo-Dionysius the Areopagite.

Pseudo-Dionysius is the most articulate representative in Western Europe of what is known as negative theology. In the East, from the Greek Fathers of the Church right up to the theology of contemporary Eastern Orthodox Christianity, apophatic theology (meaning about the same as negative theology) has been well-nigh inherent in the religious thinking of Eastern Christendom. The gist of this line of religious thought is the standpoint that God is so transcendent that effectively He is unknowable. Every time we try to describe Him, we should attach at least a mental negation to the epithets we give Him, since He is certainly not as we imagine Him.

Pseudo-Dionysius was not the first to propagate such a view. There had already been an apophatic tradition in the Eastern Church for a long time. What Pseudo-Dionysius accomplished was to turn it into a system. He was well-versed in the Neoplatonic system, in which the One is inaccessible to reason; we have a sort of access to it through its supreme emanations, the Henads. The whole of reality is a hierarchy of consecutive emanations. Pseudo-Dionysius replaced emanation, which the early Church firmly repudiated, with creation, changing the terminology to a more Christian one, putting the Choirs of Angels into the hierarchical structure, and endowing the whole concept with a strong mystical accent. According to him God is unknowable, but then our aim is not to comprehend Him but to be united with Him. Perhaps the reason why the line of thought Pseudo-Dionysius represented made such a huge impact on the theology of the West was because its hierarchical system fitted in so well with the love of organisation and classification characteristic of the Middle Ages in the West.

I shall not go into the intricacies of Pseudo-Dionysius' ideas; instead I shall cite a few sentences from an article on him,[1] which I consider worth a moment's thought at the end of this book on the search for "ultimate explanations."

> *The word "Henads" does not occur in Pseudo-Dionysius' system, instead there are names which may be attributed to God, such as Goodness, Life, Wisdom. All the time Pseudo-Dionysius insists that the Godhead Itself is beyond all these names, and It may be spoken of only in the categories of supreme negation.*

> *In Its transcendent dimension Divinity is beyond the reach of all assertion whatsoever, to such an extent that it is impossible even to assert Its existence or non-existence.*

Pseudo-Dionysius was not denying the existence of God; what he meant was to warn us against using the same language when speaking of God that we use to speak of other things. The later Scholastics would say that our language relating to God is *analogous* to our ordinary language. Pseudo-Dionysius was more radical. According to him, whenever we say anything about God – even that He exists – we are more wrong than right.

> *We say that God is the cause of all that exists, that He is the Creator. We are not so much speaking of Him and who He is, but rather who He is with respect to creation.*

We speak of the nature of God only apophatically, that is by negating all that we have said.

> *By means of relational, cataphatic names[2] it is possible to speak of God, not so much of His nature as of the way He works and of His works.*

The mainstream of Western theology did not follow the path set by Pseudo-Dionysius, however, although it never discarded the *via negativa* as one of the important paths. The Scholastic struggles with language to describe God were a good exercise preparing the way for the scientific method, but they made Western philosophers and theologians subconsciously confident that by strictly adhering to the rigours of logic they would be able to cross all the barriers. In a certain sense we are in agreement with Pseudo-Dionysius. He was not promoting irrationalism: he was not saying that the truth was whatever anyone wanted it to be, or that it all depended on psychology and social relations, or that "one opinion is as good as another." He was very far from such claims. All he was saying was that human rationality is limited. And that this reservation was a necessary condition for any human to be truly rational. That is why at the end of this book on the search for "ultimate explanations of the universe" I decided on an encounter with Pseudo-Dionysius the Areopagite.

I do not intend to go into a discussion of the relative merits of the apophatic and cataphatic trends in theology. Banking on an exclusively negative line of thought would no doubt run the risk of bringing everything to a halt; but ignoring it completely would be naïve and an oversimplification. The point is that it is not just a purely theological issue. We will find aspects of the tension between the apophatic and cataphatic styles – *toute proportion gardée* – in all thinking which reaches beyond the rigid bounds of empiricism. Especially in thinking which endeavours to face up to the task of understanding the universe.

Do we not encounter essentially the same philosophical motives in the reservations the analytical philosophers of language had about the sense of asking why something exists rather than nothing, that made Pseudo-Dionysius claim that the names we ascribe to God are merely attributions for our own ideas of what He in any case is not? Is not calling the existence of anything whatsoever (therefore also of the universe) a "brute fact for which no explanations should be sought" like the notion latent in all of Pseudo-Dionysius' reflections that the existence of the Unnameable is an "irreducible given," the grounds for everything else – in other words something of a "brute fact" as well?

Both of these opinions were a result of the same thing: a profound awareness of the most fundamental limitation of human rationality. But there is an important difference between them. The former opinion, the modern view, rules out whatever might be beyond the confirmed bounds of human cognition (in other words, it holds that whatever is beyond those bounds makes no sense). Thus it assumes that reality is geared to our potential for cognition. The latter opinion, represented perhaps somewhat haphazardly by Pseudo-Dionysius, effectively recognises the same limits to human cognition, but has an open attitude to those limits; although our knowledge of what lies beyond them is merely negative (apophatic), nonetheless it is a knowledge. The former opinion disavows the

Mystery, on the strength of its own decree as the criterion of what has a sense and what has no sense; the latter opinion immerses itself in the Mystery. The former fulfils a therapeutic function, eliminating the discomfort of ultimate questions; the latter intensifies that discomfort in order to find a remedy therein (like a vaccine which relies on the injection of viruses to make the vaccinated organism build up its immunity to them).

Pseudo-Dionysius's strategy, appropriately modified and transferred to the realm of the philosophy of science, has one more advantage in comparison with the contemporary therapeutic measures. In the light of Pseudo-Dionysius's approach the scientific adventure embarked on by mankind – not only on the quest for ultimate theories but also in the more mundane research – is not a hit-and-miss contest with brute reality, but a true Adventure of Rationality.

This book grew out of a paper I delivered at a symposium dedicated to Ludwig Wittgenstein. This philosopher shares much of the attitude assumed by Pseudo-Dionysius. His *Tractatus Logico-Philosophicus* concludes with the famous thesis that "What we cannot speak about we must pass over in silence."[3] In contemporary philosophy there are about as many interpretations of Wittgenstein as there were of Pseudo-Dionysius in the Middle Ages. So while refraining from interpretation, I shall conclude by citing two passages from the last propositions of Wittgenstein's *Tractatus* (with his emphasis marks):

The sense of the world must lie outside the world. In the world everything is as it is, and everything happens as it does happen: in *it no value exists – and if it did exist, it would have no value. If there is any value that does have value, it must lie outside the whole sphere of what happens and is the case. For all that happens and is the case is accidental. What makes it non-accidental cannot lie within the world, since if it did it would itself be accidental. It must lie outside the world.* (Proposition 6.41)

It is not how *things are in the world that is mystical, but* that *it exists.* (Proposition 6.44)

We are collectively driven by a powerful yet not fully explained instinct – to understand. We would like to see everything established, proven, laid bare. We want nothing to remain without rationale, such that would remove all suspicion, all doubt, all questions. The more important an issue, the more we desire to see it clarified, stripped of all secrets, all shades of grey. Yet this longing for "ultimate explanations" is in itself immune from being the subject of an "ultimate explanation," for when we try to understand it, we are immediately faced with the following question: what does it mean "to understand"?

NOTES AND REFERENCES

CHAPTER 1

[1] Z. Hajduk, *Filozofia przyrody – Filozofia przyrodoznawstwa – Metakosmologia* [Natural Philosophy, Philosophy of the Natural Sciences, Metacosmology (in Polish)], Lublin: Towarzystwo Naukowe KUL, 2004, pp. 148–149.

[2] Ibid., p. 149.

[3] J.B. Hartle, S.W. Hawking, "Wave Function of the Universe," *Physical Review* D28, 1983, pp. 2960–2975.

[4] Wu Zhong Chao, *No-Boundary Universe*, Changsha: Hunan Science and Technology Press, 1993.

[5] If we stay on the level of methodological reflection, this expression should be understood metaphorically, with no reference intended to the question of God.

CHAPTER 2

[1] In a letter to Bentley Newton wrote that the situation could be compared to an attempt to set up an infinite number of needles standing on their tips on the surface of a mirror.

[2] W. Thomson, *Mathematical and Physical Papers*, Vol. 2, Cambridge: Cambridge University Press, 1884, pp. 37–38.

[3] A.M. Clerke, *The System of the Stars*, London: Longmans and Green, 1890, p. 368.

[4] A. Einstein, "Kosmologische Betrachtungen zur allgemeinen Relativitätstheorie," *Sitzungsberichte der Preussischen Akademie der Wissenschaften* 1, 1917, 142–152.

5   English translation of Einstein's original paper in: *The Principle of Relativity - A Collection of Original Papers on the Special and General Theory of Relativity*, Dover, 1923, pp. 177–188; quotation from p. 180.

6   The quotation comes from *Ethics*, Spinoza's chief work, translated and edited by E. Curley, *The Collected Works of Spinoza*, Vol. 1, Princeton: Princeton University Press, 1985.

7   For Einstein's philosophical views see *Albert Einstein Philosopher-Scientist* (Library of Living Philosophers), ed. P.A. Schilpp, Chicago: Open Court, 1973; "Einstein's Philosophy of Science," *Stanford Encyclopedia of Philosophy*, http://plato.stanford.edu.

8   W. de Sitter, "On the Relativity of Inertia: Remarks Concerning Einstein's Latest Hypothesis," *Koninklijke Akademie van Wetenshappen te Amsterdam* 19, 1917, pp. 1217–1225.

9   G. Lemaître, "Note on de Sitter's Universe," *Journal of Mathematics and Physics* 4, 1925, pp. 37–41.

10 On de Sitter's world model see H. Kragh, *Cosmology and Controversy*, Princeton: Princeton University Press, 1996, pp. 11–12; P. Kerszberg, *The Invented Universe: The Einstein-De Sitter Controversy (1916–1917) and the Rise of Relativistic Cosmology*, New York: Oxford University Press, 1989.

11 A.A. Friedman, "Über die Krümmung des Raumes," *Zeitschrift für Physik* 11, 1922, pp. 377–386; idem, "Über die Möglichkeit einer Welt mit konstanter negativer Krümmung des Raumes," *Zeitschrift für Physik* 21, 1924, pp. 326–332.

12 E. Hubble, "A Relation Between Distance and Radial Velocity among Extra-Galactic Nebulae," *Proceedings of the National Academy of Sciences* 15, 1929, pp. 168–173.

13 G. Lemaître, "Un Univers homogène de masse constante et de rayon croissnat rendant compte de la vitesse radiale des nébuleuses extra-galactiques," *Annales de la Société Scientifique de Bruxelles* 47, 1927, pp. 29–39.

CHAPTER 3

1   G. Lemaître, "Rencontre avec A. Einstein," *Revue des Questions Scientifiques* 129, 1958, pp. 129–132.

2   The actual value of the cosmological constant in the real world should be determined by experimental data.

3   All the time we are talking of the dust filled cosmological models, viz. ones with an equation of state $p = 0$, where p is the pressure exerted by the "galactic gas."

4   H. Poincaré, "Sur le problème des trois corps et les equations de la dynamique," *Acta Mathematica* 13, 1890, pp. 1–270.

5   We assume that it is a Hamiltonian system confined in a "finite box" and has a finite energy.

[6] Evolution in the sense of a one-parameter mapping.

[7] In the Lebesgue sense.

[8] W. Thomson, *Mathematical and Physical Papers*, Vol. 2, Cambridge: Cambridge University Press, 1884, pp. 37–38.

[9] H.V. Helmholz, *Science and Culture. Popular and Philosophical Essays*, ed. D. Cahan, Chicago: Chicago University Press, 1995, p. 30.

[10] See my article, "Zagadnienia kosmologiczne przed Einsteinem," [Cosmological Issues before Einstein (in Polish)] *Zagadnienia Filozoficzne w Nauce* 37, 2005, pp. 32–40.

[11] L. Boltzmann, "On Certain Questions of the Theory of Gases," *Nature* 51, 1895, pp. 483–485.

[12] R. Tolman, *Relativity, Thermodynamics and Cosmology*, Oxford: Clarendon Press, 1934.

[13] Cf. S. Weinberg, *Gravitation and Cosmology*, New York: John Wiley and Sons, 1972, pp. 55–57, 593–594.

[14] Cf. M. Heller, M. Szydłowski, "Tolman's Cosmological Models," *Astrophysics and Space Science* 90, 1983, pp. 327–335.

[15] F. Tipler, "General Relativity and the Eternal Return," *Essays in General Relativity*, New York & tc.: Academic Press, 1980, pp. 21–37.

[16] More precisely, if it contains two Cauchy global planes isometric with respect to their initial conditions.

[17] But on some additional assumptions relating to the topology of the set of initial data.

[18] Technically: a compact Cauchy plane.

[19] G. Lemaître, "L'univers en expansion," *Annales de la Société Scientifique de Bruxelles* A53, 1933, pp. 51–85.

[20] Cf. the monograph on the subject: S.W. Hawking, G.F.R. Ellis, *The Large-Scale Structure of Space-Time*, Cambridge: Cambridge University Press, 1973. We shall return to the singularity problem in Chap. 18.

[21] W.J.M. Rankine, "On the Reconcentration of the Mechanical Energy of the Universe," *Philosophical Magazine* 4, 1853, p. 358. Reprinted in *Miscellaneous Scientific Papers*, London: Charles Griffin and Company, 1881, pp. 200–202.

CHAPTER 4

[1] Translation of a sentence in W. Szczerba, *Koncepcja wiecznego powrotu w myśli wczesnochrześcijańskiej* [The Concept of Eternal Return in Early Christian Thought (in Polish)], Monografie FNP, Wrocław: Wydawnictwo Uniwersytetu Wrocławskiego, 2001, p. 83.

[2] Translation of sentences quoted from p. 373 of Z. Zawirski, "Wieczne powroty światów – Badania historyczno-krytyczne nad doktryną 'wiecznego powrotu,'" [Eternal Returns of

the Worlds: A Historical and Critical Examination of the Doctrine of Eternal Returns (in Polish)], *Kwartalnik Filozoficzny* 5, 1927, pp. 328–420. This article was the first part of a series of publications on the subject (Part 2: 5, 1927, pp. 421–446; Part 2: 6, 1928, pp. 1–25).

3   C. Lanczos, "Über eine stationäre Kosmologie im Sinne der Einsteinischen Gravitationstheorie," *Zeitschrift für Physik* 21, 1924, p. 73.

4   W.J. van Stockum, "The Gravitational Field of Distribution of Particles Rotating around an Axis of Symmetry," *Proceedings of the Royal Society, Edinburgh* A57, 1937, p. 135.

5   K. Gödel, "An Example of a New Type of Cosmological Solution of Einstein's Field Equations of Gravitation," *Reviews of Modern Physics* 21, 1949, pp. 447–500.

6   A global analysis of Gödel's solution may be found in S.W. Hawking, G.F.R. Ellis, *The Large-Scale Structure of Space-Time*, Cambridge: Cambridge University Press, 1973, pp. 168–q170.

7   Closed space-like curves also occur in Gödel's solution.

8   Space-time in Gödel's solution is geodesically complete.

9   K. Gödel, "A Remark about the Relationship between Relativity Theory and Idealistic Philosophy." *Albert Einstein: Philosopher-Scientist*, ed. P.A. Schilpp. New York: Harper and Row, 1949, pp. 557–562.

10   http://www.en.wikipedia/org/wiki/Godel_metric

11   For a review and philosophical analysis of them, see J. Earman, *Bangs, Crunches, Whimpers, and Shrieks*, New York and Oxford: Oxford University Press, 1995, especially Chap. 6.

12   A discussion of these axioms is to be found in H. Mehlberg, *Time, Causality, and the Quantum Theory*, Vol. 1, Dordrecht, Boston and London: Reidel, 1980.

13   R.W. Bass, L. Witten, "Remark on Cosmological Models," *Review of Modern Physics* 29, 1957, pp. 452–453.

14   B. Carter, "Causal Structure of Space-Time," *General Relativity and Gravitation* 1, 1971, pp. 349–391.

15   J. Richard Gott III, Li-Xin-Li, "Can the Universe Create Itself?" *Physical Review* D58, 1998, pp. 23501–23543.

16   These papers concerned the polarisation of a vacuum in space-times with closed time-like curves.

17   Viz. Cauchy conditions, on the basis of which future history may be deduced.

18   See Chap. 6 below, and also my book *Granice kosmosu i kosmologii* [The Boundaries of the Cosmos and Cosmology (in Polish)], Warszawa: Scholar, 2005, Chap. 23.

19   Cf. footnote 14 in this chapter.

20   Or more strictly: no arbitrarily small disturbance of the space-time metric.

21   And monotonically.

22   My wristwatch marks out a periodic function along the time-like curve which is my history, since it shows the same time twice in every 24 h, but it would be enough to fit it

out with a 24-h clock face (or digital clock) and a calendar showing the date to obtain a function which increases at a constant (monotonic) rate.

[23] It was proved by S.W. Hawking in "The Existence of Cosmic Time Functions," *Proceedings of the Royal Society, London* A308, 1968, pp. 433–435.

[24] More precisely, the components of the metric tensor determine the structure of space-time, and at the same time are interpreted as the gravitational field potentials.

CHAPTER 5

[1] It is of no consequence that here it is a question of a "retro-prediction."

[2] F. Hoyle, "A New Model for the Expanding Universe," *Monthly Notices of the Royal Astronomical Society* 108, 1948, pp. 372–382.

[3] H. Bondi, T. Gold, "The Steady-State Theory of the Expanding Universe," *Monthly Notices of the Royal Astronomical Society* 108, 1948, pp. 252–270.

[4] The story of the emergence of the steady-state cosmology is told by Helge Kragh in his monograph on the controversy between relativistic cosmology and steady-state cosmology: H. Kragh, *Cosmology and Controversy*, Princeton: Princeton University Press, 1996, pp. 173–179.

[5] H. Bondi, *Cosmology*, Cambridge: Cambridge University Press, 1960, p. 140.

[6] Ibid., p. 141.

[7] Ibid., pp. 142–143. Bondi failed to observe that in a contracting universe it would be possible to prevent equilibrium by assuming a "continuous annihilation of matter."

[8] Ibid., p. 144.

[9] Ibid., p. 143.

[10] The reasoning is as follows: the square of the curvature of space is a measurable magnitude, e.g. via the dependence of the number of galaxies per unit volume on distance; therefore in line with the steady-state assumption it cannot change with time, while the curvature of space in an expanding universe must depend on time. The only solution to this dilemma is to assume that the curvature of space is zero.

[11] Only evolution at an exponential rate gives a steady state.

[12] F. Hoyle, *The Nature of the Universe*, Oxford: Blackwell, 1950.

[13] H. Kragh, op. cit.

[14] Ibid., p. 193.

[15] Quoted from F. Hoyle, *The Nature of the Universe*, after Kragh, op. cit., p. 192.

[16] Cf. H. Kragh, op. cit., p. 192.

[17] Quoted after Kragh, op. cit., p. 193.

[18] More on this subject in M. Heller, *Granice kosmosu i kosmologii* [The Boundaries of the Cosmos and Cosmology (in Polish)], Warszawa: Scholar, 2005, Chap. 16. See also

A. Liddle, *An Introduction to Modern Cosmology*, Chichester: John Wiley, 1999, Chap. 11.

[19] For more on this subject see W.T. Sullivan, "The Entry of Radio Astronomy into Cosmology: Radio Stars and Martin Ryle's 2C Survey," *Modern Cosmology in Retrospect*, ed. B. Bertotti, R. Balbinot, S. Bergia, A. Messina, Cambridge: Cambridge University Press, 1990, pp. 309–330; P. Scheuer, "Radio Source Counts," ibid., pp. 331–362.

[20] See M. Schmidt, "The Discovery of Quasars," *Modern Cosmology in Retrospect*, pp. 347–354.

[21] H. Kragh, op. cit., p. 343.

[22] I gave an account of the history of the discovery of this radiation in *Granice kosmosu i kosmologii*, Chap. 19. See R.W. Wilson, "Discovery of the Cosmic Microwave Background," *Modern Cosmology in Retrospect*, pp. 291–307.

[23] H. Kragh, op. cit., p. 373.

[24] Cf. ibid., pp. 358–368.

[25] The philosopher of science interested in the "invalidation" of the steady-state cosmology by the discovery of background radiation should consult T.M. Sierotowicz, *Mikrofalowe promieniowanie tła jako experimentum crucis w kosmologii?* [The Microwave Background Radiation as the Experimentum Crucis in Cosmology? (in Polish)], Kraków: Wydział Filozoficzny Towarzystwa Jezusowego w Krakowie, 1993, where the rivalry between the steady-state and relativistic cosmologies is viewed as a rivalry of research programmes in Lakatos' sense.

[26] See, for instance, M. Heller, Z. Klimek, L. Suszycki, "Imperfect Fluid Friedmannian Cosmology," *Astrophysics and Space Science* 20, 1973, pp. 205–212.

[27] Cf. M. Heller, M. Ostrowski, A. Woszczyna, "Steady-State Versus Viscous Cosmology," *Astrophysics and Space Science* 87, 1982, pp. 425–433.

[28] Incidentally, cosmological models entailing volume viscosity later entered the theory of superstrings, in view of which some modifications have been proposed for the "mechanism of viscosity." See D. Pavón, J. Bafaluy, D. Jou, "Causal Friedmann-Robertson-Walker Cosmology," *Classical and Quantum Gravity* 8, 1991, pp. 347–360.

CHAPTER 6

[1] The reader will find a straightforward but precise account of the horizon problem in A. Liddle, *An Introduction to Modern Cosmology*, Chichester: John Wiley, 1999, Chap. 12.1.2

[2] For the flatness problem, see ibid., Chap. 12.1.1.

[3] As well as several other problems with which standard cosmology has not been able to cope, which we shall not go into here.

[4] A.H. Guth, "The Inflationary Universe: A Possible Solution to the Horizon and Flatness Problem," *Physical Review* D23, 1981, pp. 347–356.

[5] A.D. Linde, "A New Inflationary Scenario: A Possible Solution of the Horizon, Flatness, Homogeneity, Isotropy and Primordial Monopole Problems," *Physics Letters* 108B, 1982, pp. 389–393.

[6] A. Albrecht, P.J. Steinhardt, "Cosmology for Grand Unified Theories with Radiatively Induced Symmetry Breaking," *Physical Review Letters* 48, 1982, pp. 1220–1223.

[7] A.D. Linde, "Chaotic Inflation," *Physics Letters* 129B, 1983, pp. 117–181.

[8] Hawking queried Linde's argumentation, pointing out that the scheme Linde proposed was not invariant but essentially depended on the resolution of space-time into time and momentary spaces; see S. Hawking, "Cosmology from the Top Down," *Universe or Multiverse?* ed. B. Carr, Cambridge: Cambridge University Press, 2007, pp. 91–98.

[9] See below, Chap. 9.

[10] We should bear in mind that energy density = energy/volume; hence in an increasing volume energy density can only remain constant if energy is accruing.

[11] This claim has been put forward by Albrecht and Steinhardt in the article cited.

[12] Readers interested in the diverse inflationary models may refer to *Inflationary Cosmology*, ed. L.F. Abbott, Singapore: So-Young-Pi, World Scientific, 1986, which contains all the original publications on inflation.

[13] G. McCabe, "The Structure and Interpretation of Cosmology: Part II. The Concept of Creation in Inflation and Quantum Cosmology," *Studies in History and Philosophy of Modern Physics* 36, 2005, pp. 67–102.

CHAPTER 7

[1] See A.H. Guth, *The Inflationary Universe. The Quest for a New Theory of Cosmic Origins*, Perseus, 1997, Chap. 17.

[2] The energy of the gravitational field is negative, since work has to be performed to separate two pieces of gravitational matter from each other.

[3] C.J. Isham, "Quantum Theories of the Creation of the Universe," *Quantum Cosmology and the Laws of Nature*, eds. R.J. Russell, N. Murphy, C.J. Isham, Vatican Observatory Publications – Berkeley and the Vatican City State: The Center for Theology and the Natural Sciences, 1993, pp. 49–89; quoted from pp. 56–57.

[4] R. Brout, F. Englert, E. Gunzig, "The Creation of the Universe as a Quantum Phenomenon," *Annals of Physics* 115, 1978, pp. 78–106.

[5] Cf. E.P. Tryon, "Cosmic Inflation," *The Encyclopedia of Physical Science and Technology*, Vol. 3, New York: Academic Press, pp. 537–571.

[6] J.B. Hartle, S.W. Hawking, "Wave Function of the Universe," *Physical Review* D28, 1983, pp. 2960–2975.

[7] Viz. the Riemann metric, which defines the geometry for a given 3-dimensional space.

[8] This description is very simplified. In fact what is meant is not the actual value of the wave function, but the square of its module.

[9] Thanks to this the integration is performed over 4-dimensional compact Riemannian spaces, which prevents the integrals from being divergent.

[10] Moreover, Hartle and Hawking seem to identify compactness of space with the absence of boundaries. However, these concepts do not overlap: a compact space may or may not have a boundary.

[11] G. McCabe, "The Structure and Interpretation of Cosmology: Part II. The Concept of Creation in Inflation and Quantum Cosmology," *Studies in History and Philosophy of Modern Physics* 36, 2005, pp. 77–78.

CHAPTER 8

[1] Examples of such relationships are to be found in Chap. 7 of Hermann Bondi's *Cosmology*, Cambridge: Cambridge University Press, 1960.

[2] Viz. the inverse of the Hubble constant.

[3] To obtain this result all the magnitudes have to be expressed in units for which the speed of light $c = 1$. This implies 1 s = 1 cm.

[4] P.A.M. Dirac, "The Cosmological Constants," *Nature* 139, 1937, p. 323; idem, "A New Basis for Cosmology," *Proceedings of the Royal Society, London* A165, 1938, pp. 199–208.

[5] R.H. Dicke, "Dirac's Cosmology and Mach's Principle," *Nature* 192, 1961, pp. 440–441.

[6] B. Carter, "Large Number Coincidences and the Anthropic Principle in Cosmology," *Confrontation of Cosmological Theories and Observational Data* (IAU Symposium), ed. M. Longair, Dordrecht: Reidel, 1974, pp. 281–289.

[7] All the other cosmological tests work on the same principle. For example, we observe microwave background radiation, therefore we reject all the cosmological models which do not permit this phenomenon.

[8] Lee Smolin wrote in *The Life of the Cosmos*, Oxford: Oxford University Press, 1997, p. 203: "The strong form [of the anthropic principle] is explicitly a religious rather than a scientific idea. It asserts that the world was created by a god with exactly the right laws so that intelligent life could exist."

CHAPTER 9

[1] L. Smolin, *The Life of the Cosmos*, New York and Oxford: Oxford University Press, 1997.

[2] Ibid., p. 93.

[3] Ibid., p. 94.

[4] Ibid., p. 96.

[5] K. Popper, *Unended Quest: An Intellectual Autobiography*, Glasgow: William Collins, 1976, p. 168.

[6] Ibid.

[7] Cf. A. Łomnicki, "Czy darwinowska teoria ewolucji jest falsyfikowalną teorią naukową ?" [Is Darwin's Theory of Evolution a Falsifiable Scientific Theory? (in Polish)], *Otwarta nauka i jej zwolennicy*, eds. M. Heller, J. Urbaniec, Kraków & Tarnów: OBI – Biblos, 1996, pp. 11–21. This article stirred up a heated discussion which may be traced in "Czy darwinizm jest metafizycznym programem badawczym czy teorią naukową ?" [Is Darwinism a Metaphysical Research Programme or a Scientific Theory – in Polish], *Zagadnienia Filozoficzne w Nauce* 22, 1998, pp. 93–113.

[8] G. McCabe, "A Critique of Cosmological Natural Selection," (2004) http://philsci-archive.pitt.edu/archive/00001648/01/NaturalSelection.pdf p. 3.

[9] Ibid. p. 9. My own critique of Smolin's concept to a large extent follows McCabe's paper.

[10] Cf. R. Penrose, "Before the Big Bang: An Outrageous New Perspective and Its Implications," *Proceedings of EPAC* 2006, pp. 2759–2762.

[11] Cf. R. Penrose, *The Emperor's New Mind: Concerning Computers, Minds, and the Laws of Physics*, Oxford and New York: Oxford University Press, 1989, p. 354.

[12] M. Tegmark, "Parallel Universes," *Science and Ultimate Reality*, eds. J.D. Barrow, P.C.W. Davies, C.L. Harper, Cambridge: Cambridge University Press, 2004, pp. 459–491; quotation from p. 465.

[13] He could start with A.F. Chalmers, *What Is This Thing Called Science?* Brisbane: University of Queensland Press, 1976 (revised edition 1999).

CHAPTER 10

[1] More, with more technical details, on structure deformation in C. Roger, "Déformations algébraïques et applications à la physique," *Gazette des Mathematiciens* no. 49, juin 1991, pp. 75–94.

[2] J. Barrow, F. Tipler, *The Anthropic Cosmological Principle*, Oxford: Clarendon Press, 1986, p. 265.

[3] Cf. G. McCabe, *The Structure and Interpretation of the Standard Model*, Amsterdam, Boston etc: Elsevier, 2007; especially Chap. 2.6.

CHAPTER 11

[1] For Wheeler's concept see J.D. Barrow, F.J. Tipler, *The Anthropic Cosmological Principle*, Oxford: Clarendon Press, 1986, pp. 369–471.

² S. Hawking, "Cosmology from the Top Down," *Universe or Multiverse?* Ed. B Carr, Cambridge: Cambridge University Press, 2007, pp. 91–98. See also a presentation for the general reader, A. Gefter, "Exploring Stephen Hawking's Flexiverse," *New Scientist*, No. 2548, 20 April 2006.

³ A. Gefter, op. cit.

⁴ M. Tegmark, "Parallel Universes," *Scientific American*, May 2003, pp. 41–51.

⁵ Ibid., p. 41.

⁶ Tegmark reinforces this commonsensical argument by invoking the ergodic property of the distribution of initial conditions: the probability distribution on the set of initial conditions for all possible universes is the same as for diverse domains in a single universe.

⁷ M. Tegmark, "Parallel Universes," *Science and Ultimate Reality*, ed. J.B. Barrow, P.C.W. Davies, C.L. Harper, Cambridge: Cambridge University Press, 2004, pp. 459–491, quotation from p. 464.

⁸ Cf. M. Rees, *Before the Beginning*, London & Sydney: A Touchstone Book, 1998.

⁹ Cf. J. Leslie, *Universes*, London & New York: Routledge, 1989.

¹⁰ S. Roush, "Copernicus, Kant, and the Anthropic Cosmological Principles," *Studies in History and Philosophy of Modern Physics* 34, 2003, pp. 5–35.

CHAPTER 12

¹ See, for example, T.F. Bigaj, *Non-Locality and Possible Worlds*, Frankfurt, Paris &tc.: Ontos Verlag, 2006, p. 69.

² Ibid., p. 72.

³ See my book, *Theoretical Foundations of Cosmology*, Singapore & London: World Scientific, 1992, Chap. 5.

⁴ G. Ellis, "Multiverses: Description, Uniqueness and Testing," *Universe or Multiverse*, ed. B. Carr, Cambridge: Cambridge University Press, 2007, pp. 387–409.

⁵ R. Nozick, *Philosophical Explanations*, Oxford: Clarendon Press, 1981.

⁶ We don't even know whether such a multiverse is a set in the technical sense of the term. There is a danger of it being liable to the same sort of antinomies as "the set of all sub-sets" in Russell's paradox.

⁷ M. Tegmark, "The Multiverse Hierarchy," *Universe or Multiverse*, ed. B. Carr, Cambridge: Cambridge University Press, 2007, pp. 99–125; quote from p. 121.

⁸ G. Ellis, op. cit., pp. 389–393.

CHAPTER 13

[1] M. Heller, *Filozofia i Wszechświat* [Philosophy and the Universe (in Polish)], Kraków: Universitas, 2006, especially in Part 2.

CHAPTER 14

[1] *Genesis* was the book's title in the Greek translation of the Old Testament known as the Septuagint, which was known already to the Jewish philosopher Aristobulus in the second century B.C.

[2] Evidence for this comes in the distinctly mnemonic form of the first chapter of *Genesis*.

[3] There is, of course, a vast body of literature on the exegetics of Chap. 1 of Genesis. In this chapter I shall be more interested in general reflection on the issue of creation rather than technical commentary. In this respect it is worthwhile reading Ernan McMullin's article, "Natural Science and Belief in a Creator: Historical Notes," *Physics, Philosophy and Theology: A Common Quest for Understanding*, eds. R.J. Russel, W.R. Stoeger, G.V. Coyne, Vatican City State: Vatican Observatory, 1988, pp. 48–79. My "unrepresentative sample" from the deluge of literature on the subject also includes H. Renckens, *Urgeschichte und Heilsgeschichte. Israels Schau in die Vergangenheit*, Mainz: Matthias Grünewald Verlag, 1961; C. Westermann, *Genesis (Biblischer Kommentar. Altes Testament 1,1)*, Neukirchen: Neukirchener Verlag, 1974, pp. 197–318; J. Ratzinger, *"In the Beginning...": A Catholic Understanding of the Story of Creation and the Fall*, Eerdmans, 1986, 1995; B.L. Bandstra, "Priestly Creation Story," *Reading the Old Testament: An Introduction to the Hebrew Bible*, Wadsworth Publishing, 1999.

[4] Cf. J. Moltmann, *God in Creation*, London: SCM Press, 1985.

[5] Biblical references in this translation come mainly from the RSV, supplemented by references to the R.C. Knox version. In this passage the R.C. Knox translation, based on the Vulgate, is closer to M. Heller's wording (except for Pol. *duch* corresponding to spirit/breath in the respective English translations but historically polysemic like the Greek *pneuma* and encompassing both meanings of the two English words), and reads: "Earth was still an empty waste, and darkness hung over the deep; but already, over its waters, stirred the breath of God." – translator's note.

[6] E. McMullin, *Evolution and Creation*, Notre Dame: University of Notre Dame Press, 1985, p.3.

[7] Ibid., p.4.

[8] For more on the subject, see my book, *Filozofia przyrody – zarys historyczny* [Natural Philosophy: A Historical Outline (in Polish)], Kraków: Znak, 2004, Chap. 2.

9   The first problem addressed by Timaios in his story are the questions, "what is it that has always existed and never known birth?" and "what is it that is always being born and never exists?" The problem concerns the Platonic distinction between the unchanging Ideas and the material world, which is only a shadow of the world of the Ideas. Even if the primeval chaos had always been in existence, it was but a shadow of the world of the Ideas. The same applies to the existence of the ordered world. However the ordered world carries more of an image of the world of the Ideas, since it was fashioned to resemble the latter, therefore it "exists more" than does chaos.

10 McMullin, op. cit., p. 6.

11 Interestingly, English versions follow the gender of the Greek noun *Logos* and use "he/ him" with reference to "the Word," whereas in Polish translations the gender of *Słowo* ("the Word") is neuter – translator's note.

12 *The Shepherd of Hermas* (J.B. Lightfoot's translation), Mandate 1 (1[26]1), http://www.earlychristianwritings.com/text/shepherd-lightfoot.html.

13 More on the subject in J. Szczerba, *Koncepcja wiecznego powrotu w myśli wczesno-chrześcijańskiej* [The Concept of Eternal Return in Early Christian Thought (in Polish)], Wrocław: Fundacja na Rzecz Nauki Polskiej, 2001, pp. 192–246. See Joseph W. Trigg, *Origen. The Bible and Philosophy in the Third Century Church*, Atlanta: John Knox Press, 1983; *Origen of Alexandria: His World and His Legacy*, eds. C. Kannengiesser, W.L. Petersen, Notre Dame: University of Notre Dame Press, 1988.

14 Ibid., p. 220. [Translation by T.B.-U.]

15 Origen, *De principiis*, I, 1, 4–5. http://www.newadvent.org/fathers/04122.htm

16 Origen, *De principiis*, II, 4. http://www.newadvent.org/fathers/04122.htm

17 Augustine, *Confessions* XI, 12 (translated by A.C. Outler) http://www.ccel.org/ccel/augustine/confessions.xiv.html

18 Ibid., IX, 14.

19 E. McMullin, op. cit, p. 2.

20 More on the subject in McMullin, op. cit, pp. 11–16.

CHAPTER 15

1   Interesting information on this subject may be found in Z. Liana, *Koncepcja Logosu i natury w Szkole w Chartres. Historyczne funkcje chrześcijańskiej koncepcji Logosu w kształtowaniu się nowożytnego pojęcia natury* [The Concept of Logos and Nature in the Chartres School: Historical Functions of the Christian Idea of Logos in the Development of the Modern Concept of Nature (in Polish)], Kraków: OBI, 1996.

² For more on the "medieval crisis" see M. Heller, Z. Liana, J. Mączka, W. Skoczny, *Nauki przyrodnicze a teologia: Konflikt i współistnienie* [Theology and the Natural Sciences: Conflict and Coexistence (in Polish)], Kraków and Tarnów: OBI and Biblos, 2001, Chaps. 4–6.

³ *In Libros Aristotelis De Caelo et Mundo*, lib. I, 1, 7, n.6

⁴ E.g. in the *Summa Contra Gentiles*, Chaps. 15–33.

⁵ To denote the concept of contradiction St. Thomas uses the Latin expression *repugnantia intellectui*: "things which are repugnant to the intellect." The evidence that this phrase is synonymous for Thomas with self-contraction comes in the sentence ... *propter repugnantiam intellectuum aliquid non posse fieri, sicut quod non potest fieri ut affirmatio et negatio sint simul vera* ... ("that something cannot hold due to 'repugnance of intellects' [self-contradiction], just as affirmation and negation cannot hold true at the same time" – *De Aeternitate Mundi*, n. 2)

⁶ *Productio rei secundum totam eius substantiam.*

⁷ *Non enim ponitur, si creatura semper fuit, ut in aliquo tempore nihil sit, sed ponamus quod natura eius talis esset si sibi reliqueretur* – *De Aeternitate Mundi*, n. 7

⁸ Ibid., n. 8.

CHAPTER 16

¹ A. Funkenstein, *Theology and the Scientific Imagination from the Middle Ages to the Seventeenth Century*, Princeton: Princeton University Press, 1986, p. 125. This book presents an excellent study of the links between theology and the emergence of the experimental sciences. This chapter is based largely on Funkenstein's Chap. 3.

² *De Usu Partium*, after Funkenstein, op. cit., p. 125, footnote 8.

³ The term "contingent" was first used by Alexander of Hales.

⁴ Cf. A. Funkenstein, op. cit., pp. 179–192.

⁵ An expression of this was the famous "Tree of Porphyry," universally accepted throughout the Middle Ages.

⁶ E. Cassirer, *Substance and Function. Einstein's Theory of Relativity*, New York: Dover Publications, 1953.

⁷ R. Hooykaas, *Religion and the Rise of Modern Science*, Edinburgh: Scottish Academic Press, 1972.

⁸ G.W. Leibniz, *Théodicée*. English translation by E.M. Huggard. http://www.gutenberg.org/catalog/world/readfile?fk_files=206453&pageno=146

⁹ More on this subject in my book *Uchwycić przemijanie* [To Grasp Transience (in Polish)], Kraków: Znak, 1977, pp. 125–132.

CHAPTER 17

[1]   I. Newton, *Principia*, Vol. II, *The System of the World*, trans. A. Mott, ed. F. Cajori, Berkeley etc.: University of California Press, 1962, p. 546.

[2]   Cf. A. Funkenstein, *Theology and the Scientific Imagination from the Middle Ages to the Seventeenth Century*, Princeton: Princeton University Press, 1986, Chap. 3.

[3]   *Principia*, Vol. II, p. 545

[4]   Ibid.

[5]   Quoted after *Newton's Philosophy of Nature – Selections from His Writings*, ed. H.S. Thayer, New York & London: Hafner Publishing Company, 1974, pp. 38–39. In Newton's reasoning here it is easy to spot basically the same strategy applied today in the anthropic arguments.

[6]   More on this subject in D. Kubin, "Newton and the Cyclical Cosmos: Providence and the Mechanical Philosophy," *Science and Religious Belief*, ed. C.A. Russell, London: University of London Press, The Open University Press, 1973, pp. 147–169.

[7]   Cf. Domenico Bertoloni Melli, "Newton and the Leibniz-Clarke correspondence." *The Cambridge Companion to Newton*, eds. I. Bernard Cohen, George E. Smith, Cambridge: University Press, 2002. Cambridge Collections Online. Cambridge University Press. 19 March 2009 DOI:10.1017/CCOL0521651778.XML.017

[8]   In fact contemporary cosmology is facing the same problem, as we saw in Part I of this book.

CHAPTER 18

[1]   Published in *Die philosophischen Schriften von G.W. Leibniz*, ed. C.I. Gerhardt, Vol. VII, Halle, 1846–1863, pp. 190–193.

[2]   In the original the sentence reads *Cum Deus calculat et cogitationem exercet, fit mundus.*

[3]   G.W. Leibniz, *Theodicée*, English translation by E.M. Huggard
http://www.gutenberg.org/catalog/world/readfile?fk_files=206453&pageno=43

[4]   http://www.gutenberg.org/catalog/world/readfile?fk_files=206453&pageno=52

[5]   http://www.gutenberg.org/catalog/world/readfile?fk_files=206453&pageno=65

[6]   http://www.gutenberg.org/catalog/world/readfile?fk_files=206453&pageno=43

[7]   *Monadologie* 29, trans. Jonathan Bennett
http://www.earlymoderntexts.com/pdf/leibmon.pdf

[8]   *Monadologie* 31, trans. Jonathan Bennett
http://www.earlymoderntexts.com/pdf/leibmon.pdf

[9]   *Monadologie* 32, trans. Jonathan Bennett
http://www.earlymoderntexts.com/pdf/leibmon.pdf

[10] *De Contingentia*, in G.W. Leibniz, *Textes inédits*, ed. G. Grua, Paris: Presses Univs. de France, 1948, Vol. I, pp. 303–306.

[11] Monadologie 46 http://www.earlymoderntexts.com/pdf/leibmon.pdf

[12] http://www.gutenberg.org/catalog/world/readfile?fk_files=206453&pageno=78

[13] http://www.gutenberg.org/catalog/world/readfile?fk_files=206453&pageno=78

[14] http://www.gutenberg.org/catalog/world/readfile?fk_files=206453&pageno=78

[15] http://www.gutenberg.org/catalog/world/readfile?fk_files=206453&pageno=78

[16] M. Planck, "Das Prinzip der kleinsten Wirkung," *Kultur der Gegenwart*, Leipzig: B.G. Teubner, 1915.

[17] *Monadologie* 40 http://www.earlymoderntexts.com/pdf/leibmon.pdf

[18] *Monadologie* 41 http://www.earlymoderntexts.com/pdf/leibmon.pdf

[19] G.W. Leibniz, *On the True Theologia Mystica. Philosophical Papers and Letters*, ed. and trans. Leroy Loemker, Reidel, 1969, p. 368.

[20] Leibniz's third letter to Clarke, 25 February 1716, 6. Trans. Jonathan Bennett. http://www.earlymoderntexts.com/pdf/leibclar.pdf

[21] Ibid. 4.

[22] We may assume that by writing about things that are "coexistent" Leibniz meant "simultaneous." However in the above quotation he did not want to use the word "simultaneous," since simultaneity cannot have a sense until it is specified what is meant by "time." Andrzej Staruszkiewicz and I have published our reading of Leibniz's polemic with Clarke in "Polemika Leibniza z Clarke'iem w świetle współczesnej fizyki," [Leibniz's Polemic with Clarke in the Light of Modern Physics] *Wieczność, czas, kosmos* [Eternity, Time, Cosmos (in Polish)], Kraków: Znak, 1995, pp. 41–54.

[23] For more on this subject, see my book *Filozofia i Wszechświat* [Philosophy and the Universe (in Polish)], Kraków: Universitas, 2006, especially Part II.

[24] *Theodicée* http://www.gutenberg.org/catalog/world/readfile?fk_files=206453&pageno=82

[25] M. Dummett, What Is Mathematics About?" *Philosophy of Mathematics. An Anthology*, ed. D. Jacquette, Oxford: Blackwell, 2002, pp. 19–29, quoted from p. 22.

CHAPTER 19

[1] I discussed this problem more extensively in Chap. 20 of my book *Granice kosmosu i kosmologii* [The Boundaries of the Cosmos and Cosmology (in Polish)], Warszawa: Scholar, 2005. For more see R. Geroch, G.T. Horowitz, "Global Structure of Spacetime," *General Relativity. An Einstein Centenary Survey*, eds. S.W. Hawking, W. Israel, Cambridge: Cambridge University Press, 1979, pp. 212–293.

2 In the theory of relativity an object's length depends on the choice of a reference system; so we are not talking about the length of the curve but the possibility of its extension in a strictly defined sense.

3 It is assumed that no domain has been artificially removed from this space-time. This assumption is of course given a mathematical formulation.

4 At least that's what it seemed at the time when the singularity theorems were being formulated. Currently opinions on "what is physically realistic" in very early stages of the universe's evolution have undergone a significant change (see below).

5 S. Hawking, G. Ellis, *The Large-Scale Structure of Space-Time*, Cambridge: Cambridge University Press, 1973, p. 364.

6 I have written more extensively on the singularities and the latest research on them in "Cosmological Singularity and the Creation of the Universe," *Creative Tension*, Philadelphia & London: Templeton Foundation Press, 2003, pp. 79–99.

7 Which we may do only for the sake of discussion, since it can hardly be assumed that in very dense states of the universe gravitation will not manifest its quantum features.

CHAPTER 20

1 S.C. Morris. *Life's Solution. Inevitable Humans in a Lonely Universe*, Cambridge: Cambridge University Press, 2003, p. 12.

2 We made some preliminary remarks on this subject in Chap. 12.

3 I have written more extensively on this subject in Chap. 3 of my book *Filozofia i Wszechświat* [Philosophy and the Universe (in Polish)], Kraków: Universitas, 2006.

4 See, for instance, M. Rédei, S.J. Summers, "Quantum Probability Theory," *Studies in History and Philosophy of Modern Physics* 38, 2007, pp. 390–417.

5 Cf. D.V. Voiculescu, K. Dykema, A. Nica, *Free Random Variables*, CRM Monograph Series, Vol. 1, American Mathematical Society, Providence, 1992.

6 E. McMullin, "Evolutionary Contingency and Cosmic Purpose," *Studies in Science and Theology* 5, 1997, pp. 91–112; the quoted passage is on pp. 106–107.

7 Ibid.

CHAPTER 21

1 R.L. Kuhn, "Why This Universe? Toward a Taxonomy of Possible Explanations," *Skeptic* 13, No. 2, 2007, pp. 28–39.

2 Ibid., p. 28.

3 English translation © Jonathan Bennett http://www.earlymoderntexts.com/pdf/leibphg.pdf

⁴ Leibniz regards his monadology as the foundation of physics, but in fact it is a purely metaphysical doctrine.

⁵ Op. cit. 7. Parts of sentence originally stressed by Leibniz italicised in the translation.

⁶ Ibid.

⁷ Ibid., 8.

⁸ Cf. *The Stanford Encyclopedia of Philosophy*, http://plato.stanford.edu/entries/nothingness/

⁹ We don't even have a statement to say that nothing exists, for there is no-one capable of formulating such a statement.

¹⁰ Cf. footnote 19 in the cited article by R.L. Kuhn.

¹¹ However, we should not forget that mathematics is not just a language; I have written on this subject in *Filozofia i Wszechświat* [Philosophy and the Universe (in Polish)], Kraków: Universitas, 2006 (see especially Chaps. 5 and 6).

¹² W.V. Quine, "On What There Is." *From a Logical Point of View*, Harvard University Press, 1964, p. 16.

¹³ P. van Inwagen, "Why Is There Anything At All?" *Proceedings of the Aristotelian Society 70*, 1996, pp. 95–110.

¹⁴ Of course we could define the probability distribution function in the set of all universes (on condition that it exists in that set – see in Chap. 12 Sect. 3) in such a way as to define the probability of the occurrence of an empty world as zero – but that would be begging the question.

¹⁵ H. Eilstein, *Biblia w reku ateisty* (in Polish), Wydawnictwo IFiS PAN: Warszawa, 2006. The book does not appear to have been published in an English translation up to now (2009).

¹⁶ All the passages cited in this sub-chapter come from pp. 115 to 117 in Eilstein's book.

EPILOGUE

¹ J.A. Kłoczowski, "Teologia negatywna – miądzy dialektyką a mistyką," [Negative Theology: Between Dialectics and Mysticism (in Polish)] *Znak* No. 613, 2006, pp. 71–94.

² The aim of cataphatic theology, the reverse of apophatic theology, is to make a positive statement on God.

³ L. Wittgenstein, *Tractatus Logico-Philosophicus*, English translation by D.F. Pears and B.F. McGuinness, Project Gutenberg online edition http://www.gutenberg.org/dirs/etext04/tloph10.txt

INDEX

Printed in the United States
By Bookmasters